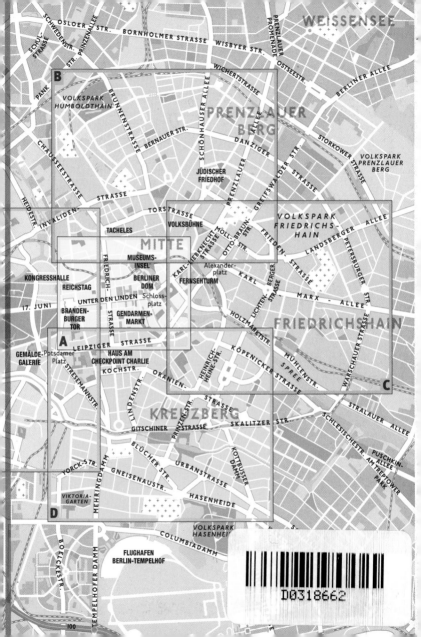

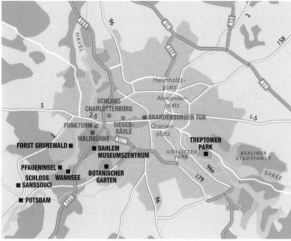

EXCURSIONS

SHOPPING

Department stores
KaDeWe (**F** F4)
Germany's answer to
Harrods: the store that
stocks everything.
Galeries Lafayette (**A** C3)
→ *Friedrichstr. 76-78*
Mon-Sat 10am–8pm
Predominantly luxury,
ready-to-wear, accessories
and perfumery.
Dussmann (**A** C2)
→ *Friedrichstr. 90*
Mon-Sat 10am–midnight
A large cultural store selling
books, CDs, German and
international DVDs.
Shopping malls
Kaufhof (**C** A2)
→ *Alexanderplatz*
Europa Center (**F** F3)
→ *Breitscheidplatz*
Potsdamer Platz
Arkaden (**E** D4)
→ *Potsdamer Platz*
Friedrichstadt
Passagen (**A** C3)
→ *Französischestr. 68*

Flea markets
Tiergarten (**E** C3)
→ *Str. of the 17. Juni*
Sat-Sun 11am–5pm
Art and flea market
covering several miles.
Am Zeughaus (**A** D2)
→ *Am Kupfegraben*
Sat-Sun 11am–5pm
The most attractive of the
flea markets, with books,
records and Eastern
European crafts.
Trodelmarkt am
fehrbelliner Platz (off **F** D4)
→ *Sat-Sun 10am–4pm*
Second-hand knickknacks.
Arkonaplatz (**B** C2)
→ *Sun 10am–5pm*
The trendiest.
Note: foreign credit cards
(Visa, AmEx etc.) are
often refused.

BERLIN
ANOTHER WAY

Berlin by bus
Lines 100 and 200
→ *Between the Zoologischer*

Garden and Alexanderplatz,
or between Prenzlauerberg
and Potsdamer Platz;
every ten mins
A tour of Berlin's most
important monuments.
Berlin by boat
Many companies provide
boat trips through the
center or along the Havel.
Reederei
Bruno Winkler
→ *Embarking from*
*Schlossbrücke (**A** E2) and*
*Reichstagufer (**A** C1)*
Tel. 349 95 95 March-Oct:
daily at 10.20am, 11am,
2.15pm, 3pm; duration: 3 hrs;
€16; with dinner Fri-Sat at
7pm; duration 2 ½ hrs; €25;
www.reedereiwinkler.de
Museuminsel, Tiergarten,
Kreuzberg.
Stern und Kreisschiffahrt
→ *Tel. 536 36 00*
www.sternundkreis.de;
from €7.50
From Wannsee Lake to
Werder Island, via the
Pfaueninsel.

EXCURSIONS

Treptower Park
→ *S-Bahn Treptower Park*
Large East Berlin park,
with a monument to the
Russian soldiers killed
during World War Two.
Botanischer Garten
→ *S-Bahn Botanischer*
Garten. Tel. 838 50 100
Daily 9am–dusk
A sumptuous 82-foot-
high glasshouse with
18,000 plant species.
Forst Grunewald
→ *S-Bahn Grunewald*
Immense forest in the
southwest of Berlin.
Wannsee
→ *S-Bahn Wannsee*
The largest lake in the
city: several miles of
beach and the
Pfaueninsel, with its
peacocks and gardens.
Dahlem
Museumszentrum
→ *Lansstr. 8, Dahlem Dorf*
subway station
Tel. 830 14 38 Tue-Fri 10am–
6pm; Sat-Sun 11am–6pm
Ethnography and non-
European arts.
Potsdam
Schloss Sanssouci
→ *Tel. (0331) 969 41 90*
Tue-Sun 10am–6pm
(5pm Nov-March)
The rococo Versailles of
Frederick the Great.
Dutch Quarter
→ *Friedrich-Eberstr.,*
Kurfürstentr., Gutenbergstr.
Built in the 18th century,
it houses art galleries,
cafés and restaurants.
Filmstudio Babelsberg
→ *Grossbeerenstr.*
Tel. (0331) 721 27 50
March-June, Sep: Tue-Thu,
Sat-Sun; July-Aug, Oct: daily
One of the largest movie
studios in Europe
between 1917 and 1945.

THE WALL

■ Aug 12, 1961: the construction of the Wall begins. It will be 96 miles long, 12 feet high with 300 watchtowers ■ 1961–89: more than 239 people are killed trying to cross the Wall ■ Nov 9, 1989: fall of the Wall.

Remains of the Wall
Bernauer Strasse (B B2)
Fragment and monument.
Potsdamer Platz (E D4)
Fragment.
Brandenburger Tor (A A3)
Red markings (2 ½ miles) on the sidewalks as far as Checkpoint Charlie.
East Side Gallery (C D4)
650-foot-long fragment.

POTSDAM

MARX AND ENGELS

HACKESCHE HÖFE

to the nearest note as you settle up (rather than leave change on the table).

Restaurants
Kneipe
Bars that serve drinks, sandwiches, light meals or more elaborate fare.
Imbiss
Snack bars (vans, bars or small restaurants) serving *Curry Wurst, Döner Kebab*, falafel, fried noodles and pizzas at a snip.
Biergarten
Gardens filled with large tables and wooden benches, and buzzing with life. Convivial atmosphere and typically German: perfect for a good beer and sausage!

Specialties
Berliner Weisse
Blonde beer with raspberry syrup.
Bulette
Spiced, minced meat ball.
Curry Wurst
Sausage with a curry and ketchup sauce.
Eisbein auf Sauerkraut:
Ham on sauerkraut (cabbage).

BARS AND CLUBS

There are many *Kneipen*, cocktail bars, *Biergarten*, clubs, cultural centers and nightclubs of all varieties. Anything goes in most clubs: the most far-out or trendiest clothes can be seen under the same roof as jeans and T-shirts. Entry fees are negotiable for groups. Admission to clubs is often free after 4am.

MUSEUMS

Information
Staatliche Museum zu Berlin
→ *Tel. 266 3660*
www.smb.spk-berlin.de
Information on the major museums.
3-Tage Karte

→ *On sale in all participating museums; €15 (€7.50 concessions)*
Transferable ticket valid for three days for admission to over 60 museums, including the largest ones.

Guided tours
Bärentouren
→ *Tel. 46 06 37 88*
Themed tours.
Individuelle Touren
→ *Tel. 892 13 38*
Individual tours.
Zeitreisen Berlin
→ *Tel. 44 02 44 50*
Historical tours.

SHOWS

Listings
Tip and Zitty
→ *Bimonthly; available from newspaper vendors; www.zitty.de*
Berlin life: movies, clubs, theaters, concerts and useful addresses.
030
→ *Monthly and bimonthly;* from bars, clubs and some theaters; www.berlino30.de
Small, free publications listing parties and concerts.

Reservations
→ *On websites such as www.deutschland-tickets.de, by phone, from the kiosks (Theaterkasse), or direct from the theaters*
Beware: the large venues are often booked out months in advance (Deutsche Oper, Philharmonie, etc.).
Hekticket (C A2)
→ *Karl-Liebknecht-Str. 12*
Tel. 24 31 24 31
www.hekticket.de
Last-minute tickets: opera, concerts, theater, etc.
Abendkasse
→ *In person one hour before the show*
Box offices are reopened and a waiting list is operated. Unclaimed tickets are re-sold 30 minutes before the performance.

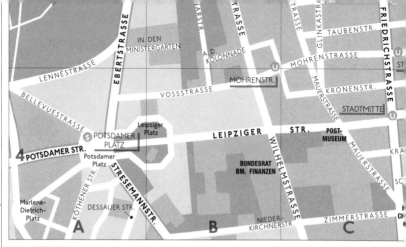

SCHINKELMUSEUM

MUSEUMSINSEL

PERGAMONMUSEUM

Brandenburger Tor (A A3)
→ *Pariser Platz*
Built in 1788 by Langhans after the model of the propylaea in Athens, the Brandenburg Gate has a Quadriga on its top. A symbol of peace and then of German nationalism, the gate ended up in East Germany when the Wall was built. Its reopening on December 22, 1989, made it an emblem of reunification.

Holocaust-Mahnmal (A B3)
→ *www.holocaust-mahnmal.de*
This project by the American architect Peter Eisenman opened in 2005 after years of deliberation. The Field of Stelae is made of 2,700 shiny black tombstones, whose simplicity makes the memorial even more moving. In the 'Raum der Namen', the names of some 700 victims of the Holocaust are projected onto the walls, while details of their lives are being heard through loudspeakers.

Unter den Linden (A D2)
With Baroque ornamentation (Zeughaus), classical colonnades (Staatsoper, Humboldt Universität), neoclassical pillars (Kronprinzenpalais, Altes Palais) and a Doric portico (Neue Wache, Schinkel's masterpiece), this avenue forms a guard of honor for the statue of Frederick the Great (1851).

Schinkelmuseum (A D3)
→ *Friedrichswerdersche Kirche, Werderstr. Tel. 208 13 23 Tue–Sun 10am–6pm*
A master of 19th-century neoclassicism in Berlin, Karl Friedrich Schinkel was also the father of architectural eclecticism, evident in the neo-Gothic brick church (1824–30) that houses his museum. The gallery is lined with sketches, manuscripts and technical directions. The nave boasts some neoclassical statues.

Gendarmenmarkt (A D3)
→ *Huguenot Museum in the Französischer Dom Daily noon (11am Sun)–5pm*
Twin domes (Von Gonthard, 1785) crown the Deutsche and Französischer Dom (1701–5), built for the Fren and German Calvinists. Since the fall of the Wall, cafés-terraces have invade the beautiful paved squar

Bebelplatz (A D2-3)
St Hedwig's Cathedral (174 73) looks like an upturned cup, and the Baroque curv of the old library are (1774 80) like a *Kommode*. Only the Staatsoper reflects the initial plans for the Forum devised by Frederick II (1740–86). In the evening, white light glows through glass paving stone at the center of the square. Belo the empty shelves of a library recall the Nazi *auto*

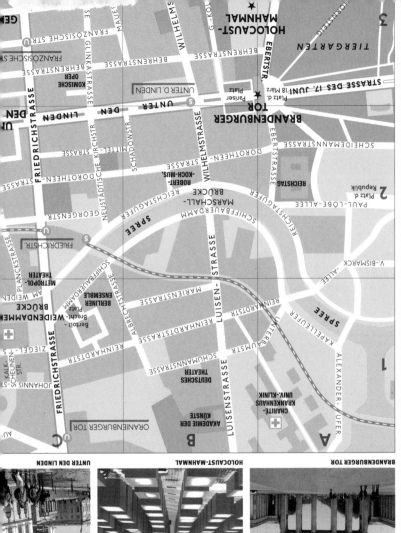

↓ Map B

BRANDENBURGER TOR

HOLOCAUST-MAHNMAL

UNTER DEN LINDEN

Unter den Linden, in the heart of historical Berlin (Mitte), is being gradually restored to its former splendor. Setting off from the Brandenburg Gate, visitors can take a trip back through two centuries of German monumentalism, from the former Soviet embassy and the contemporary buildings on Friedrichstrasse, to Frederick II's Bebelplatz. Modern, classical or Baroque, the majestic façades convey the austere grandeur of the Prussians. At the end of the avenue, opposite the site of the former royal palace, stands the Museumsinsel, the cultural Acropolis of the city popularly known as Athens-on-the-Spree.

OXYMORON

LUTTER & WEGNER

IMBISS, RESTAURANTS

Suppenbörse (A D2)
→ *Dorotheenstr. 43*
Tel. 20 45 59 03 Mon–Fri
11am–6pm; Sat noon–4pm
(6pm in summer)
Recipes from India, France, Swabia or Hungary. Every week you choose from six new and filling soups. Take out or eat in. Dishes €3–5 .

Ständige Vertretung (A B2)
→ *Schiffbauerdamm 8*
Tel. 28 59 87 36
Daily 10.30am–1am
The photos of *Wessis* ('from the west') politicians on the walls leave no room for doubt: the sympathies of the owner, from Bonn, lie with West Germany. Berlin *Buletten*, of course, but also specialties from the Rhineland. Dishes € 8–16.

Samâdhi (A B3)
→ *Wilhelmstr. 77*
Tel. 22 48 88 50
Daily noon–3pm, 6–11pm;
closed Sun dinner
What this place lacks in decor, it makes up for in the quality of its food. Oriental vegetarian cuisine, subtle flavors and very good value. One of the few restaurants in this area, where embassies are ubiquitous. Do try the ginger and lemon flavored dishes. Dishes €9–16.

Oxymoron (A E1)
→ *Rosenthalerstr. 40/41*
Tel. 28 39 18 86
Daily 11am–midnight
At the heart of Hackesche Hofe, a former garage has given way to a theatrical decor. Carpets from the 1950s, velvet sofas and sparkling chandeliers: a haven of tranquility much appreciated by Berliners. Dishes €9–21.

Französischer Hof (A D3)
→ *Jägerstr. 56*
Tel. 20 17 71 70
Daily 11am–midnight
This large brasserie is popular with chic Berliners, who monopolize the terrace in the summer to enjoy the concerts given in front of the Konzerthaus. Inside, wonderful Jugendstil decor, stylish global cuisine and a flurry of waiters swirling around in a perfect choreography. In the evenings, jazz or *Kabarett* in the piano-bar. Dishes €15–19.

Lutter & Wegner (A D3)
→ *Charlottenstr. 56*
Tel. 20 29 540 Daily 11am–
2am (service noon–midnight)
This 1920s Berlin brasserie, filled with contemporary artworks, is an institution. Do as Marlene Dietrich used to, and come here

GLOSSARY

Bahnhof: railway station
Brücke: bridge
Berg: mountain
Kirche: church
Mauer: wall
Platz: square
Schloss: palace
Strasse: street
Viertel: area

REUNIFIED BERLIN

BRD: former RFA
DDR: former RDA
Ossis: former Eastern Germans
Ostalgie: joining of *Ost* and *Nostalgie*
Wessis: former Western Germans
Wende: watershed (1989)

HAUS DER KULTUREN DER WELT / HOUSE OF WORLD CULTURE

FRESCO ON THE WALL OF THE EAST SIDE GALLERY

concert in the Waldbühne (see Excursions).
July-August
Classic open air
→ *Beg July*
Open-air classical concerts in Gendarmenmarkt (**A** D3).
Christopher Street Day
Gay march in Ku'Damm (**D**).
September
Internationales Literaturfestival
Readings and literary gatherings across the city.
Art Forum Berlin
→ *One weekend Sep-early Oct; www.art-forum-berlin.com*
International modern art fair in the Messehalle (off **F** A4).
October
Lesbian Film Festival
at the Acud (**B** C3) and SO 36 (**D** F2) movies theaters; *www.womenfilmnet.org*
November
Jazz Fest Berlin
ajor jazz festival in e Haus der berliner

Festspiele, **F** E4); *www.berlinerfestspiele.de*
December
Weihnachtsmärkte
Christmas markets.
Silvester
→ *Dec 31*
Fireworks across the city.

OPENING HOURS

Mealtimes
Restaurants tend to be open noon–3pm and 6pm–10pm. It's difficult to find somewhere to eat after 11pm...
Nightlife
...but it isn't difficult to stay up all night. There are no closing times in Berlin: the nights are long and the bars stay open until the last customer leaves.
Museums
→ *Often closed Mon (national museums) or Tue, Dec 24-26, Dec 31-Jan 1; some museums stay open late on Thu (10pm)*

Banks
→ *Mon-Fri 9am–5pm; often closed Wed and Fri am*
Post offices
→ *Mon-Fri 9am–6pm; Sat 9am–1pm*
Central post office
→ *7 Joachimstaler Str. Daily 8am–midnight*
Stamps can only be bought at the post office.
Shops
→ *Mon-Fri 10am–6pm/8pm; Sat 10am–6pm*
Sales are when the stores choose to have them.

EATING OUT

Customs
In the morning, Berliners have a large *Frühstück* with *Brötchen* slathered in jam, cooked meats and pickled herring. Then a light meal or snack at noon, and a big dinner in the evening.
Tipping
It is customary to leave a tip by rounding off the bill

ARCHITECTURE

Medieval architecture
There are very few remains from this period; **Nikolai Kirche** (**C** A2).
Baroque
Flamboyant style, which retains a certain simplicity in the North; **Unter den Linden** (**A** D2).
Rococo
Gilt work and pastel tones during Frederick II's reign; **Schloss Sanssouci, Neues Palais (Potsdam)**.
Neoclassicism
Style inspired by antiquity, represented by Schinkel; **Neue Wache** (**A** D2) **Altes Museum** (**A** E2).
The Hobrecht Plan
A plan of urban expansion (1862) based on the Haussmannian model, rendered essential by the increase in population during the 19th century; **Savignyplatz** (**F** D3).
Wilhelminian style (or **Gründerzeit**). Heteroclite style with strong indus-trial resonances. Marked a new era of expansion, born of the creation of the German Empire in 1871.
Modern architecture
Unbridled creativity under Weimar (1920s), particularly with the New Objectivity, the transition toward functionalism and linear forms; **Museum für Gestaltung** (**F** B4).
Socialist architecture
Grandiose walkways and somber buildings (East Berlin, after 1950); **Karl-Marx-Allée** (**C** B-E2).
Modern architecture
Since 1989, Berlin has become a melting pot of the greatest architects in the world; **Potsdamer Platz** (**E** D4).

BERLIN
Everyman MapGuides

Welcome to Berlin!

This opening fold-out contains a general map of Berlin to help you visualise the six districts discussed in this guide, and four pages of valuable information, handy tips and useful addresses.

Discover Berlin through six districts and six maps

A Unter den Linden / Friedrichstadt / Museumsinsel

B Scheunenviertel / Prenzlauer Berg

C Alexanderplatz / Nikolaiviertel / Friedrichshain

D Kreuzberg

E Tiergarten / Potsdamer Platz / Schöneberg

F Kurfürstendamm / Charlottenburg

For each district there is a double-page of addresses (restaurants – listed in ascending order of price – cafés, bars, tearooms, music venues and shops), followed by a fold-out map for the relevant area with the essential places to see (indicated on the map by a star ★). These places are by no means all that Berlin has to offer, but to us they are unmissable. The grid-referencing system (**A** B2) makes it easy for you to pinpoint addresses quickly on the map.

Transportation and hotels in Berlin

The last fold-out consists of a transportation map and four pages of practical information that include a selection of hotels.

Thematic index

Lists all the street names, sites and addresses featured in this guide.

TADSHIKISCHE TEESTUBE

BERLINER ENSEMBLE

FRIEDRICHSTADT PASSAGEN

for the best of Austrian and German cooking, with a view of the Gendarmenmarkt. Dishes €16–23.

TEAROOMS

Opernpalais (A D2)
→ Unter den Linden 5
Tel. 20 26 83
Daily 8am–midnight
Smart elderly ladies, trendy young Berliners, and women in evening dresses gather beneath the moldings and gilt work of the Prinzessinnen Palace. The terrace is popular for afternoon tea or a drink before the opera.

Tadshikische Teestube (A D2)
→ Am Festungsgraben 1
Tel. 204 11 12 Daily 5pm (3pm Sat-Sun)–midnight
A journey to Tadjikistan awaits he who enters this house: a veritable artisanal gem filled with wooden sculptures and paintings. Come and take off your shoes before relaxing on the rugs and cushions. Reserve to avoid disappointment.

CLASSICAL MUSIC, THEATERS

Staatsoper (A D2)
→ Unter den Linden 7
Tel. 20 35 45 55
Ticket office: Mon-Fri 11am–7pm; Sat-Sun & public hols 2–7pm and one hour before performance;
www.staatsoper-berlin.org
The oldest opera house in the world, commissioned by Frederick II, was built in 1741 as a classical temple dedicated to music. Wonderful recitals and operas are staged here.

Komische Oper (A C3)
→ Behrenstr. 55-57
Tel. 20 26 06 66
Ticket office: Mon-Sat 11am–7pm; Sun 1–4pm and one hour before performance;
www.komische-oper-berlin.com
This is an opera house for everyone: great classics of comic opera and works performed in German.

Konzerthaus (A D3)
→ Gendarmenmarkt 2
Tel. 20 30 92 101
Ticket office: Mon-Sat noon–7pm; Sun noon–4pm;
www.konzerthaus.de
Hung with chandeliers, Schinkel's auditorium has superb acoustics for the prestigious Konzerthaus orchestra. Look out for the exceptional organs at the rear of the stage.

Berliner Ensemble (A C1)
→ Bertolt-Brecht-Platz 1
Tel. 28 40 81 55
Ticket office: Mon-Fri 8am–

6pm; Sat-Sun 11am–6pm;
www.berliner-ensemble.de
The theater of Bertolt Brecht, then of Heiner Müller, is now run by Claus Peymann. Works by the great masters as well as by contemporary writers.

Deutsches Theater (A B1)
→ Schumannstr. 13a
Tel. 28 44 12 25
Ticket office: Mon-Sat 11am–6.30pm; Sun 3–6.30pm
Berlin's oldest theater, made famous by Max Reinhardt when he ran it in the 1920s. Today, Thomas Langhof mounts revivals of great classics.

BAR, CLUB

Cinéma Café (A E1)
→ Rosenthalerstr. 39
Tel. 28 06 415 Mon-Fri 5pm–3am; Sat-Sun noon–3am
A very long, narrow café bustling with people day and night. Find a spot in which to drink your beer: at the piano, next to the window or at a table on the sidewalk. Terrace in the courtyard in summer.

Kalkscheune (A C1)
→ Johannisstr. 2
Tel. 59 00 43 40
www.kalkscheune.de
A vast complex of rooms for diverse events: rock or jazz concerts, cabaret evenings, dance or disco.

SHOPPING

Berlin Story (A B3)
→ Unter den Linden 10
Tel. 20 45 38 42
Daily 10am–7pm
Everything to do with Berlin, from keyrings to specialist books. At the rear is a huge model of the classical city and the royal palace, now gone.

Friedrichstadt Passagen (A C3)
→ Friedrichstr. 68
Mon-Sat 10am–8pm (6pm Sat)
A sumptuous and labyrinthine shopping mall, whose stores underground are connected by passageways.

Lutter & Wegner (A D3)
→ Charlottenstr. 56
Tel. 20 29 540
Daily 11am–midnight
A grand store offering a guided tour of the tastes of the German charcuterie: cured meats, sausages and black pudding washed down with a glass of wine, to enjoy here or to take out.

Grüne Erde (A D1)
→ Oranienburgerstr. 1-3
Tel. 20 45 59 03 Mon-Sat 11am (10am Sat)–8pm
A temple dedicated to well-being! Everything sold here is made of 100 percent natural ingredients for you and your home.

GENDARMENMARKT

BEBELPLATZ

Map

D

STRASSE
✚ ST. HEDWIG-
KRANKENHAUS
NEUE
NIENBURGERSTR. SYNAGOGE
KRAUSNICKSTRASSE

🅢
ANIENBURGER STR.

E ✈
TUCHOLSKY- ZIEGELSTR.
STR.

MONBIJOUSTR.

GESCHWISTER- AM KUPFERGRABEN
SCHOLL-STR.

ORGENSTRASSE AM KUPFERGRABEN

UNIVERSITÄTSSTR. DOROTHEENSTRASSE

HUMBOLDT
UNIVERSITÄT

★ UNTER DEN LINDEN
EN ALTES
PALAIS STAATSOPER

BEBELPLATZ

EHRENSTRASSE

RANZÖSISCHE STRASSE

FRANZÖSISCHER
✚ DOM
RMENMARKT
★

E

GR. HAMBURGER STR.

SOPHIENSTR.
SOPHIEN-
KIRCHE

HACKESCHE
HÖFE

Monbijou-
pl.

ORANIENBURGERSTRASSE

NEUE
PROMENADE

MONBIJOU PARK

🅢
HACKESCHER MARKT

BODEMUSEUM

PERGAMONMUSEUM

ALTE
NATIONALGALERIE
FRIEDRICHS-
BRÜCKE

★ BURGSTR.
MUSEUMSINSEL
ALTES
MUSEUM
BODESTRASSE AM LUSTGARTEN

AM
FESTUNGSGRABEN ZEUGHAUS
(DEUTSCHES
HISTORISCHES
NEUE MUSEUM)
WACHE SCHLOSS-
BRÜCKE

★ KRONPRINZENPALAIS

**SCHINKEL-
MUSEUM**
★
ST.-HEDWIGS

FRIEDRICHS-
WERDERSCHE
KIRCHE

LUSTGARTEN

**BERLINER
DOM**
★

**DDR
MUSEUM**

LIEBKNECHT-
BRÜCKE

SCHLOSS-
PLATZ
PALAST DER
REPUBLIK

WERDERSCHE

UNTER

WASSERSTR.

KURSTR.

OBERWA

JÄGERSTRASSE

F

MAX-BEE
STRASSE

WEINMEISTERSTR. 🅤

NEUE SCHÖNHAUSER MÜNZSTRASSE
STRASSE
ROSENTHALERSTR.

ROCH-
STR. STRASSE

MITTE

DIRCKSENSTRASSE

ROCHSTRASSE

KARL-LIEBKNECHT-STRASSE

FERNSEHTURM

MARIENKIRCHE

RATHAUSSTR.

ROTES
RATHAUS

NIKOLAI-
KIRCHE

STADT-
BIBLIOTHEK

POSTSTRASSE

RATHAUSSTR.

Molken-
markt

NIKOLAIVIERTEL

GRUNER-
STR.

MÜHLENDAMM
BRÜCKE

DAMM

SPREE

MÜHLEN-

SPANDAUER STRASSE

SPANDAUER STRASSE

BURGSTR.

1

2

↑ Map C

BREITE STRASSE

BRUDERSTRA

FRIE

SE FI

3

FISCHER-
INSEL

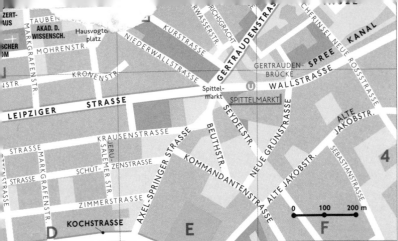

Map D →

...SEUMSINSEL / BODEMUSEUM

ALTES MUSEUM

ALTES MUSEUM

...fe of May 10, 1933, when ...,000 'anti-German' books ...ent up in smoke below the ...niversity's windows.

...useumsinsel (A E2)
→ Tel. 20 90 55 77
...aily 10am–6pm (10pm Thu);
...ww.museumsinsel-berlin.de
...n amazing complex of
...useums built like many
...mples to the grandeur of
...russia (1830–1930).

...odemuseum
...opped with an impressive
...upola, the neo-Baroque
...uilding houses fabulous
...orks of art from the
...yzantine period and
...ntiquity, and an
...mpressive coin collection.

...ergamonmuseum
...eparating the two sides of

this vast museum stands
the enormous Pergamon
altar. To the left, Hellenistic
and Latin statues and
mosaics. To the right, the
market gate from Miletus
(120 BC) opens onto the
blue and gold archway of
the Ishtar Gate, which
marks the entrance to the
department of Near-Eastern
antiquities and Islamic art.

Altes Museum
Resembling a Greek temple,
Schinkel's impressive
building houses one of the
finest ancient collections in
the world: statues from
Ancient Greece to the late
Roman era, vases, jewelry
and gold and silver crockery.
Until 2009 the Altes Museum

will exhibit the collections
from the Ägyptisches
(Egyptian) Museum, of
which the bust of Queen
Nefertiti (c. 1340 BC),
Akhenaton's beautiful wife,
is one of the highlights.

Alte Nationalgalerie
→ Tue–Sun 10am–6pm
(10pm Thu)
The best-known German
and European 19th-century
artists are represented here:
Von Menzel, Caspar D.
Friedrich, Manet, etc.

Berliner Dom (A E2)
→ Lustgarten; Tel. 202 690
Mon–Sat 9am–7pm;
Sun noon–5pm
Beneath the opulent nave
of the Protestant cathedral,
rebuilt in 1905 in a rather

overpowering neo-Baroque
style, are 90 Hohenzollern
tombs (ceremonial coffins
of Frederick I and Sophie-
Charlotte in the style of
Schlüter, and that of the
Great Elector).

DDR Museum (A E2)
→ Karl-Libknechtstrasse
Tel. 847 123 73
Daily 10am–8pm (10pm Sat);
www.ddr-museum.de
A multimedia museum
that attempts to give an
impression of what life was
like in Socialist Germany.
Visitors can rummage
through the cupboards of
a typical kitchen, watch East
German TV and even drive
a Trabi through a virtual
Soviet-style housing estate.

MUSEUM FÜR NATURKUNDE

GEDENKSTÄTTE BERLINER MAUER

★ Hackesche Höfe (B C4)
→ Rosenthalerstr. 41 / Sophienstr. 6
The façades designed in 1906 by Kurt Berndt and August Endell, whose style was similar to that of the Viennese Sezession, transformed this labyrinth of inner courtyards into a masterpiece of Jugendstil: ceramic tiles, vibrant colors and an interplay of curves and geometric figures. Completely restored, they now house popular cafés, a movie theater, a theater, galleries and showrooms.

★ Sophienstrasse (B C4)
A traditional narrow street typical of living conditions in the 18th and 19th centuries: behind the stuccoed façades is a sequence of brick courtyards. At no. 18 you will see the neo-Renaissance earthenware portico of the Artisans' House, a key site in Communist history. Nearby stands the Baroque church tower of the Sophienkirche (1712–34).

★ Neue Synagoge (B C4)
→ Oranienburgerstr. 28-30
Tel. 88 02 83 00 May-Aug: Sun-Fri 10am–6pm; (8pm Sun-Mon; 5pm Fri); Sep-April: Sun-Fri 10am–6pm (2pm Fri);
This Byzantine dome sparkling in the sunshine belongs to the great synagogue (1866), torched during Kristallnacht, then bombed. Behind the Moorish arches of the façade, rebuilt in 1995, a museum examines the role of the Jewish community in Berlin.

★ Tacheles (B B4)
→ Oranienburgerstr. 54/56
Tel. 282 61 85
This artistic squat, a symbol of alternative culture, has managed to survive the property developers and tourist coaches since 1990, and finally the council has agreed to help the area retain its artistic ambience. Studios, a movie theater and an auditorium have been built in the immens[e] building covered with frescos and graffiti. To the right of the entrance, the steel dragon of the Zapata Café belches fire over customers' heads.

★ Brechthaus (B A3)
→ Chausseestr. 125
Tel. 283 05 70 44 Guided to[urs] every 30 mins; Tue-Fri 10am– noon; Thu 10am–noon, 5– 6.30pm; Sat 9.30am–1.30pm; Sun 11am–6pm (every hour); restaurant: Tue-Sat 6–9pm; www.lfbrecht.de
This was the home of Bertolt Brecht and Helene Weigel from 1953 to 1956. In the basement, the Kelle[r] restaurant serves Viennese

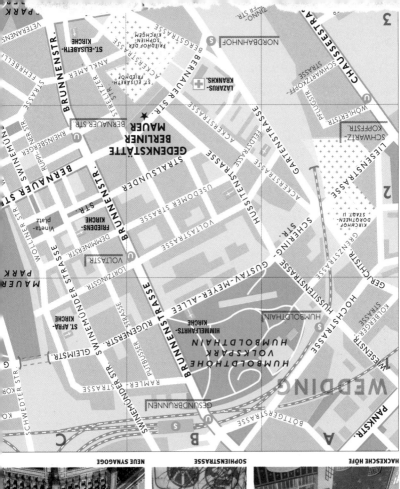

NEUE SYNAGOGE

SOPHIENSTRASSE

HACKESCHE HÖFE

Scheunenviertel / Prenzlauer Berg

North of Mitte, the Scheunenviertel ('barn district'), with its galleries, lofts and bars, has become a trendy bohemian area since reunification. It is hard to believe that before World War Two these restored stucco façades concealed squalid Mietskasernen, home to the communities which were then regarded as 'lower classes': the laborers, crooks and the Jewish community. Further out, in Prenzlauer Berg, renovated buildings stand next to peeling walls, which are gradually disappearing, while bars and alternative arts centers are opening their doors to young people in search of cheap and trendy ways to relax in the evenings.

ZOSCH

SCHWARZE PUMPE

IMBISS, RESTAURANTS

Konnopke (**B** E2)
→ *Schönhauser Allee 44a*
Mon-Fri 6am–8pm;
Sat 11.30am–7pm
The oldest *Curry Wurst* in the city, run by the same family since the 1930s. Dishes €1.25–3.50.

Gorki Park (**B** D3)
→ *Weinbergsweg 25*
Daily 10am–2am
Hip retro feel with a 1950s interior. Enjoy Russian specialties such as salad laced with vodka dressing late into the night. Dishes €3.50–12.

Zosch (**B** B4)
→ *Tucholskystr. 30*
Daily 4pm–2am
A Mitte classic: generous omelets for intimate evenings for two or live ska music in the basement. Dishes €5–10.

Frida Kahlo (**B** E1)
→ *Lychener Str. 37*
Tel. 445 70 16 Daily 10am–2am (3am Fri-Sat)
Inspired by Mexico's *Casa Azul* (museum dedicated to Frida Kahlo), the blue façade of this restaurant is instantly recognizable. Good Mexican cuisine and popular brunch. Terrace in summer. Dishes €5–12.

Miro (**B** E1)
→ *Raumerstr. 29*

Tel. 44 73 30 13
Daily 10am–midnight
Dimly lit Anatolian restaurant, with bare brick walls and wooden floors, warm service and friendly staff. Grilled meats, mezze, and surprisingly good Turkish and Anatolian wines. Dishes €5–15.

Walden (**B** D2)
→ *Choriner Str. 35*
Tel. 44 900 25   Mon-Sat from 5pm; Sun from 10am (5pm July-Aug)
A candlelit restaurant with homemade food – even down to the bread and the pasta. Don't miss out on the potato cakes with fresh pan-fried salmon and spicy avocado mousse. Dishes €5.50–12.

CAFÉS, CLUBS

Beth Café (**B** B4)
→ *Tucholskystr. 40*
Tel. 281 31 35
Mon-Fri 11am–5pm (3pm Fri)
Kosher food, right beside the synagogue.

Nemo (**B** D2)
→ *Oderberger Str. 46*
Daily 6pm–5am
Surrounded by a growing retinue of trendy bars, the Kapitän Nemo has held the helm since the Wall came down. Lively conversation, chinking glasses and laughter: a

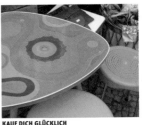

◄LKSBÜHNE **ACUD** **KAUF DICH GLÜCKLICH**

sociable melee that the comic-strip paintings, eccentric decor and confusion of irregularly shaped tables fortunately do nothing to subdue.

Schwarze Pumpe (B D3)
→ *Choriner Str. 76*
Tel. 449 69 39
Daily 10am–1am
At the heart of the LSD Viertel, this *Kneipe* owes its name to the petrol pump that is the central feature of the room. Big wooden tables at which students come to have breakfast or sit and read.

Wohnzimmer (B E1)
→ *Lettestr. 6; Tel. 445 54 58*
Daily 10am–5am
This 'living room' has been in the wars: broken-down couches and armchairs, dilapidated walls and lighting suitable for a blackout. People have been partying here every night for almost six years. If you want to chill out on the sofas, you need to get here in the afternoon.

Schlot Kunstfabrik (B B3)
→ *Schlegelstr. 29*
Tel. 448 21 60 Daily from 8pm; concerts from 9pm
A café offering musical entertainment in the basement of a former factory: soft lighting and no more than 20 tables.

Modern jazz (Thu-Mon), *Kabarett* (Tue, Sun afternoon).

Prater (B D2)
→ *Kastanienallee 7-9*
Tel. 448 56 88 Daily noon–midnight; restaurant: Mon-Sat 6pm–midnight; Sun noon–midnight
Pleasant *Biergarten* under the plane trees of a peaceful courtyard; but Prater is also an annex of the Volksbühne, a club playing house music, and a *Gaststätte* (restaurant) serving slightly pricey but very good German cuisine (brunch on Sun).

THEATERS, CULTURAL CENTERS

Volksbühne (B D4)
→ *Rosa-Luxemburg-Platz*
Tel. 24 06 57 77 Opening times vary according to venue; www.volksbuehne-berlin.de
This former People's Theater is still avant-garde and politically engaged. The two auditoria are certainly eclectic in their programming: salsa, tango and ballroom dancing (Grüner Salon, Wed-Fri 9pm); readings and songs (Roter Salon). At the weekend both become clubs, sometimes taking over the entire theater.

Acud (B C3)
→ *Veteranenstr. 21*
Tel. 449 10 67
www.acud.de
Threatened with closure, this cultural center has been saved by the locals. Behind the façade now undergoing renovation, there is art and music on every floor: reggae, bossa or hip-hop concerts in the café, theater, gallery, art movies and a club. Ideal for getting to know the more alternative residents of the district.

Kulturbrauerei (B E2)
→ *Knaackstr. 97*
Tel. 48 49 44 44
Behind the brick walls of a lavishly restored industrial brewery is an upmarket and very trendy complex: theaters, *Kneipen*, art galleries, three nightclubs (packed on Sat) and a multiplex movie theater.

SHOPPING

Jenny Paris (B B4)
→ *Linienstr. 141*
Tel. 27 59 40 34
Wed-Fri 2–7pm; Sat 2–4pm
In this tiny boutique, you will find a talented jewelry designer whose specialty is hair clips made of silver.

Tribaltools.de (B E2)
→ *Lychener Str. 10*
Tel. 35 10 21 03 Mon-Sat

noon–8pm (4pm Sat);
www.tribaltools.de
The hub of Berlin's trance scene: records, CDs, bags, tribal T-shirts and DJs mixing music.

Weinerei (B C3)
→ *Veteranenstr. 14*
Tel. 440 69 83 Mon-Fri 1–8pm; Sat 11am–8pm
1,001 wines explained by an experienced sommelier who entrances the locals when he does wine tasting evenings.

Frau Tulpe (B C3)
→ *Veteranenstr. 19*
Tel. 44 32 78 65 Mon-Fri 10am–8pm; Sat noon–6pm
Tables for cutting fabric to order and rolls of fabric to choose from: welcome to the studio of Frau Tulpe, who creates 1970s-style bags and kitsch accessories.

Kauf Dich Glücklich (B D2)
→ *Oderberger Str. 44*
Tel. 44 35 21 82 Mon-Fri 11am–1am (2am Fri); Sat-Sun 10am–1am (2am Sat)
In this street formerly divided by the Wall is this unusual café looking more like a bazaar. And indeed, whether the bric-à-brac or the clothes, everything is for sale. Trendy and not always particularly good value, but bargain hunters flock here.

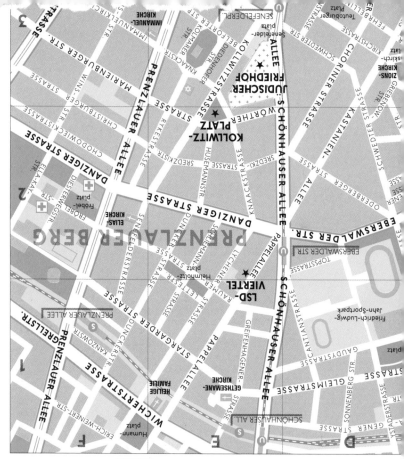

BRECHTHAUS

CHELES

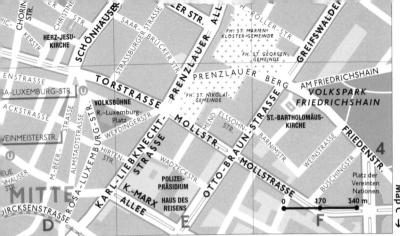

...SCHER FRIEDHOF

KOLLWITZPLATZ

LSD VIERTEL

ipes once cooked by
lene. The house is next to
e Dorotheenstädtischer
edhof, where Brecht and
igel, Hegel, Schinkel,
ler and Heinrich Mann
e buried.

**Museum für
turkunde (B** A3)
→ *Invalidenstr. 43*
20 93 85 91
t-Sun 10am–6pm;
e-Fri 9.30am–5pm;
ww.museum.hu-berlin.de
ildren visiting the old
atural History Museum
889) are fascinated by
e impressive dinosaur
eletons: the 72-foot-long
achiosaurus is the
gest skeleton of a reptile

in the world.

**★ Jüdischer
Friedhof (B** E3)
→ *Schönhauser Allee
Mon-Thu 8am–4pm;
Fri 8am–1pm*
Under a tall archway of
trees, ivy-covered tombs
lie beneath dead leaves.
The weather is gradually
eroding the names of Max
Liebermann and Meyerbeer,
as well as the scars inflicted
by the Nazis. The synagogue
at 53 Rykerstr. was spared
the flames on Kristallnacht
(Nov 9, 1938).

**★ Gedenkstätte
Berliner Mauer (B** B2)
→ *Dokumentationszentrum:
Bernauer Str. 111; Tel. 464 10 30*

*Tue-Sun 10am–5pm (6pm
summer)*
An immaculate section of
the Wall stands between
two huge sheets of metal,
seemingly frozen forever in
time. The information center
opposite charts the history
of the 'Antifascist Protection
Wall' erected in 1961 by the
DDR to prevent the mass
exodus of its inhabitants.

★ Kollwitzplatz (B E2)
This pretty square at the
heart of Prenzlauer Berg is
less popular with the
alternative Berlin scene
since the façades regained
their stucco and pastel
hues, but it is still enjoyable
to sip a Berliner Weisse on

the terrace or in one of
the cafés on Husemannstr.
At the center of the square
stands the statue of
Käthe Kollwitz.

★ LSD Viertel (B E2)
An old bastion of
underground culture in
Prenzlauer Berg, its
nickname is taken from the
initials of its three main
streets: Lychener Str.,
Schliemannstr. and
Dunckerstr. The alternating
pattern of restored façades
and dilapidated buildings
reflects the profile of its
inhabitants, who are either
advocates of a quiet life or
young aficionados of the
countless bars in the area.

FERNSEHTURM

ALEXANDERPLATZ

VOLKSPARK FRIEDRICHSHAIN

★ Märkisches Museum (C B3)

→ Am Köllnischen Park 5
Tel. 30 86 60 Tue, Thu-Sun 10am–6pm; Wed noon–8pm
Although this building resembles a Gothic monastery, it was built in 1908 to house a museum. Its maze of rooms and corridors retrace the history and civilization of Berlin, from the first traces of prehistoric settlers to everyday life in the 19th century, including Gothic sculptures from the Marienkirche. Don't miss the stereoscopic photos of the Kaiserpanorama, dating from the late 19th century.

★ Marx-Engels-Forum (C A2)

→ Karl-Liebknecht-Str., Spandauer Str., Rathausstr.
Inaugurated in 1986 by Erich Honecker, this monument is dedicated to the 'fathers of socialism' – Marx and Engels – who are throwing a paternal glance toward the East. Nearby, a fresco celebrates the joys of socialism and decries the capitalist world.

★ Nikolaiviertel (C A2)

A maze of little streets and medieval houses dating back to 1987! The tiny St Nicholas district was completely reconstructed (sometimes using concrete)

around the Nikolaikirche to celebrate the 750th anniversary of the foundation of Berlin. The Gothic nave of the oldest monument in the city (1230–1878) houses an annex of the Märkisches Museum. Nearby, the rococo balconies of the Ephraïm Palace (banker to Frederick II) are a reconstruction.

★ Rotes Rathaus (C A2)

→ Spandauer Str. 100
The brick-colored silhouette of the 'Red City Hall' (1869), an example of architectural eclecticism, draws its inspiration from Italian Romanesque: its belfry was

modeled on the campanile built by Giotto for the Duomo in Florence. It is surrounded by a terracotta frieze that charts the history of Berlin up to the foundation of the Empire.

★ Marienkirche (C A2)

→ Spandauer Str. 100
Tel 242 44 67
Daily 10am–9pm (4pm winter tours by appt, in German
One of Berlin's few remaining medieval buildings, St Mary's church pioneered the use of Brandenburg brick (end of 13th c.; pinnacle turret from 1418). Inside, the stark white walls, the rhythmic harmony of the vaults and the light

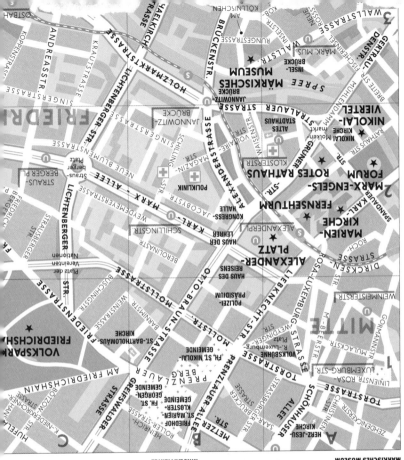

NIKOLAIVIERTEL

MÄRKISCHES MUSEUM

The 1,200-foot-high television tower, a symbol of the former DDR, soars above a patchwork of architectural styles that reflect the successive political stances: the vast perspectives of the Stalinian Karl-Marx-Allee and the modernist buildings on Alexanderplatz, which stand next to the small studios in the St Nicholas district; this was entirely rebuilt in 1987. To the east, the gaily colored façades of the punk squats in Friedrichshain, with their alternative bars and clubs, and low rents, are turning Simon-Dach-Strasse into the new hub of the Berlin *Szene*.

SEASON

UMSPANNWERK OST

RESTAURANTS

Papaya (C F3)
→ *Krossenerstr. 11*
Tel. 29 77 12 31
Daily noon–midnight
Thai cuisine made with extra fresh ingredients flown directly from Bangkok. Very reasonable prices. Dishes €6–10.

Sauerkraut und Bulgur (C D3)
→ *Strasse der Pariser Kommune 35*
Tel. 29 77 36 31
Daily 11am–midnight
A blend of Mediterranean and German tradition? A daring gamble... a winning gamble for these two young patrons bursting with ideas. Their restaurant stands out in sharp contrast to the solemnity of Karl-Marx-Allee. Dishes €6–10.

Lafil (C A1)
→ *Gormannstr. 22*
Tel. 285 990 26
Mon-Fri 1pm–1am;
Sat-Sun 6pm–1am
A taverna in Basque and Spanish colors that's always packed. Renowned for its *pinchos*, little skewers of meat and grilled vegetables, which are so tasty you don't need to be hungry to gorge on them. Dishes €7–19.

Umspannwerk Ost (C D2)
→ *Palisadenstr. 48*
Tel. 42 08 93 23
Daily 11.30am–midnight
This former factory now provides the setting for nouvelle cuisine. The dishes are served beneath steel vaults, delighting both your tastebuds and your wallet. The red tagliatelle with chicken, mushrooms and white wine sauce, are absolutely divine. Dishes €9–14.

Zur letzten Instanz (C B2)
→ *Waisenstr. 14-16*
Tel. 242 55 28 Daily noon
(11pm Sun)–1am
The oldest *Kneipe* in Berlin (dating back to 1621) still has the remarkable heated seat on which Napoleon once sat when he lunched here. The restaurant serves a solid traditional style of cuisine: platter of assorted meats (with sauerkraut and *Kartoffeln*), or *Buletten*. Dishes €9–16; set menu €18–22.

Alt-Berliner Wirtshaus Henne (C B4)
→ *Leuschnerdamm 25*
Tel. 614 77 30
Tue-Sun 7pm–midnight
Numerous specialties of roast chicken

ACHKAMMER

DIE TAGUNG

KAUFBAR

suggested by the restaurant's name. Rustic setting and terrace in summer. À la carte €12.

CAFÉS, BARS

Café Schönbrunn (C D1)
→ *Am Friedrichshain*
Tel. 46 79 38 93
Daily 10pm–1am
By day a plain café-restaurant in the middle of Volkspark, where customers of all ages come to eat or take a break. By night a rather more fashionable bar with lounge music and a laid-back, youngish ambience.

Astro Bar (C E3)
→ *Simon-Dach-Str. 40*
Daily from 8pm
Plastic astronauts and screens straight out of *Star Trek*: a cocktail bar in a spaceship. To withstand take-off you can sink into deep, low sofas, which make coming back down to earth a little tricky.

Die Tagung (C F4)
→ *Wühlischstr. 29*
Tel. 29 77 37 88
Daily from 7pm
Ostalgie, red flags, busts of Lenin, official portraits and ensigns from the Communist Youth Movements are enjoying renewed popularity in Berlin, sometimes through

genuine nostalgia, but often in mockery. The Tagung serves Roter Oktober, the beer once drunk by the crack corps of the Red Army. There's a club in the basement.

Dachkammer (C E3)
→ *Simon-Dach-Str. 39*
Tel. 296 16 73 Mon-Fri noon–3am (4am Fri); Sat-Sun 10am–3am (4am Sat)
A quieter, friendly *Kneipe*: good for a quiet chat over a drink. Wide selection of teas, all available in low-caffeine varieties.

CLUBS, CONCERTS

Fischladen (C F2)
→ *Rigaer Str. 83*
Café: daily 5pm–3am; club: Sat from 10pm
It can take over 15 minutes to walk through the four tiny rooms of this café-club, packed until 4am: two bars with outrageous decor, a reggae dance floor and another in the cellar, which is more of a dance-hall. Last but not least: the minuscule punk bar, where a hip crowd drinks beer, listening to reggae or trash-punk.

Maria am Ostbahnhof (C C3)
→ *Schillingbrücke*
Tel. 21 23 81 90 Wed-Sat from 11pm; concerts during

the week (from 8 or 9pm); *www.clubmaria.de*
This is the club for the techno avant-garde. Excellent music and internationally renowned DJs.

Berghain (C D3)
→ *Wriezenerstr.*
Wed-Sat from midnight; www.berghain.de
One of the top spots for Berlin's 'technophiles'.

SHOPPING

c/o Atle Gerhardsen (C C3)
→ *Holtzmarktstr. 15-18*
Tel. 65 41 83 41
Tue-Sat 11am–6pm; www.atlegerhardsen.com
This art gallery is worth a visit, less to buy, perhaps, than to witness the furious creative energy of modern Berlin. The exhibitions of installations and paintings are changed regularly.

Naturkost Friedrichshain (C F3)
→ *Boxhagener Str. 109*
Tel. 29 66 04 52 Mon-Sat 8.30am–8pm (6pm Sat)
This organic supermarket has a very wide choice of produce at prices that even beat the major supermarkets. The secret of its success, however, is its bread counter with

amazing fresh produce.

Prachtmädchen (C F4)
→ *Wühlischstr. 28*
Tel. 97 00 27 80 Mon-Sat 11am–8pm (4pm Sat)
With 1930s decor for a chic and engaging style (mainly in muddy purples and sky blue colors). Striped tights and flowery boots: it's spring all year here.

Kaufbar (C F4)
→ *Gärtnerstr. 4*
Tel. 29 77 88 25
Daily 10am–1am (closed Tue in summer)
People come here to drink a cup of tea, and end up buying the cup. This is more than just a bar: it's a bar-cum-second-hand shop, where everything that you see here can either be consumed or purchased.

Chapati Design (C E3)
→ *Simon-Dach-Str. 37-38*
Tel. 29 04 40 23
Mon-Sat 11am–8pm (7pm Sat)
On the left, a completely handmade ready-to-wear fashion in a hippie chic style that displays an Indian inspiration: velvet flares, gold-embroidered shawls. On the right, crafts imported direct from Azerbaijan and the Maghreb: Persian rugs, kilims and glassware.

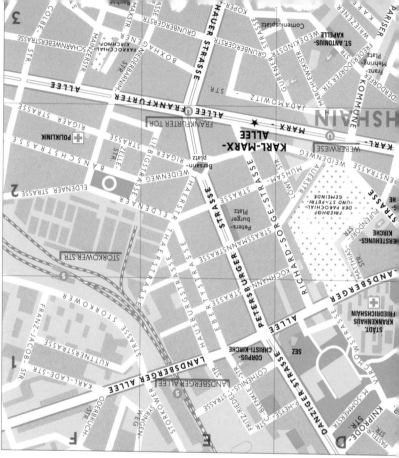

MARIENKIRCHE

ROTES RATHAUS

MARX-ENGELS-FORUM

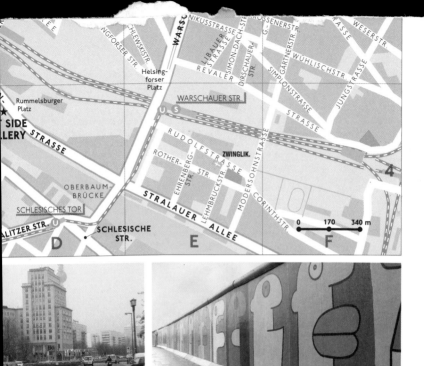

Map labels (top portion):

NIKLUSSTRASSE · WARSCHAUER · LIBAUER STRASSE · SIMON-DACH-STR. · ROSSENERSTR · GARTNERSTR · WESERSTR · STRASSE · WÜHLISCHSTR · JUNGSTRASSE

INGFORSER STR · HILEWSKISTR · REVALER · DIRSCHAUER STR. · SIMPLONSTRASSE

Helsing-forser Platz · WARSCHAUER STR. · STRASSE

Rummelsburger Platz · SIDE GALLERY · STRASSE

RUDOLFSTRASSE · ZWINGLIK.

ROTHER- · EHRENBERG- STR · LEHMBRUCKSTR · MODERSOHNSTRASSE · CORINTHSTR

OBERBAUM-BRÜCKE · STRALAUER ALLEE

SCHLESISCHES TOR · SCHLESISCHE STR.

ALITZER STR.

0   170   340 m

D   E   F

4

ARL-MARX-ALLEE

EAST SIDE GALLERY

---

treaming through the
indows enhance the
xuberant fittings: the
aroque pulpit by Andreas
chlüter (1703) and the
rgan by J. Wagner (recitals
t 4.30pm on Sat, May-Oct).
o the left of the entrance
a 15th-century Dance
f Death.

★ Fernsehturm (C A2)
Tel 242 33 33 Daily 9am–
idnight (10am–midnight in
ov-Feb); last ascent 11.30pm
isible from all of Berlin, the
200-foot-high spire of the
elevision Tower, built in
969, defined West
ermany from its dizzy
eights. There is a fabulous
iew from the panoramic

platform and the revolving
restaurant, housed in the
steel sphere at 666 feet.

★ Alexanderplatz (C A2)
Despite its gigantic
proportions, the 'Alex' is still
the true hub of East Berlin.
Completely bombed, the
working-class counterpart of
Potsdamer Platz made way,
in the 1960s, for a huge
esplanade surrounded by
concrete blocks. At the
center stands the World
Time Clock by Erich John,
and a fountain of the
Friendships of Peoples.

**★ Volkspark
Friedrichshain (C** C1)
Behind the neo-Baroque
Fairy Tale fountain (1913)

and its statues taken from
Grimm, stretch the woods
and grass of the People's
Park (1840). In summer,
Berliners flock here to
sunbathe. In winter, the two
hills, made from debris from
a bombed bunker and local
landfill, are used as
toboggan runs. This is the
burial place of 200 victims
of the 1848 revolution.

★ Karl-Marx-Allee (C E2)
The asphalt of the former
Stalineallee is tank-proof:
the massive parades from
the East used to rumble up
this colossal avenue to
reach Alexanderplatz. The
building of the historicist-
Stalinian buildings on this

'Moscow-on-the-Spree'
gave rise, on June 17, 1953,
to a workers' revolt, which
was violently put down to
prevent it spreading
throughout the city.

★ East Side Gallery (C D4)
→ Mühlenstr.
www.eastsidegallery.com
Stretching for over 4,265
feet, this is the world's
longest outdoor art gallery.
In 1990, 106 artists used the
East side of the Wall as a
vast blank page. Fresco after
fresco has created a
cavalcade of colors.
Damaged by the weather
and by tourists who try to
pull away fragments, they
are slowly being restored.

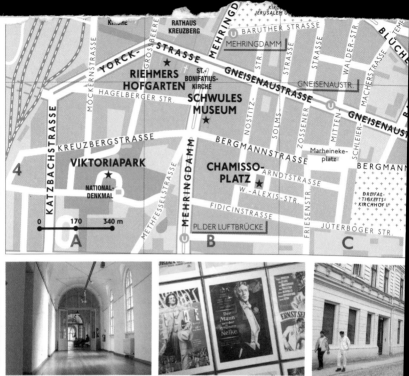

KÜNSTLERHAUS BETHANIEN

SCHWULES MUSEUM

RIEHMERS HOFGARTEN

★ Haus am Checkpoint Charlie (D B1)

→ Friedrichstr. 43-45
Daily 9am–10pm

On the site of the crossing point between the American and Soviet sectors, this museum presents a chronological history of the Wall with accounts by defectors and the amazing stratagems some used successfully to escape, such as hiding in suitcases. On the top floor, there is an exhibition on non-violence.

★ Stiftung Topographie des Terrors (D B1)

→ Niederkirchnerstr. 8
Tel. 254 86 703 Daily 10am–8pm (6pm Oct-April)

The sinister headquarters

of the Third Reich (Gestapo, secret services, Waffen SS) stood on this piece of wasteland until 1945. The foundations of underground cells, uncovered in 1987, now house an exhibition on Nazism and deportation.

★ Martin Gropius Bau (D A1)

→ Niederkirchnerstr. 7
Tel. 25 48 60
Wed-Mon 10am–8pm

When building the School of Applied Arts (1887–91), Schmiede and Gropius took a leaf out of Schinkel's book: neo-Renaissance eclecticism and graceful decoration (ceramics and mosaics). The building is now used for major exhibitions.

★ Deutsches Technikmuseum (D A3)

→ Trebbinerstrasse 9
Tel. 90 25 40 Tue-Fri 9am–5.30pm; Fri-Sat 10am–6pm;
www.dtmb.de

The already gigantic Museum of Technology has gained a new wing, where the aeronautics and space collections are now exhibited. Also, don't miss the Spectrum science center with its dozens of hands-on experiments explaining the workings of acoustics, optics, electricity, and more.

★ Jüdisches Museum (D C2)

→ Lindenstr. 9-14; Tel. 25 993
300 Daily 10am– 8pm (10pm
Mon); www.jmberlin.de

The symbolic, 'deconstructivist' structure of this building by Daniel Libeskind (1999) is shaped like an exploded Jewish star and attracted hundreds of thousands of visitors long before any exhibits were brought inside the museum. The focus of the permanent exhibition is on the historical role of the Jewish community right up to its annihilation by the Nazis.

★ Künstlerhaus Bethanien (D F2)

→ Mariannenplatz 2; Tel. 616
90 30 Wed-Sun 2–7pm

This is an art center in a 19th-century hospital, where, leading off the huge neo-Romanesque foyer,

D

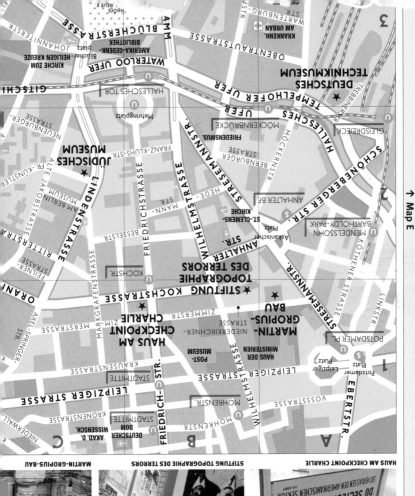

MARTIN-GROPIUS-BAU

STIFTUNG TOPOGRAPHIE DES TERRORS

HAUS AM CHECKPOINT CHARLIE

Flanked by the Wall on two sides and therefore abandoned by the *Wessis* investors, Kreuzberg was the haunt of dropouts, Turkish laborers and young rebels until 1989. Punks and pacifists flocked here from all over West Germany to avoid military service, and they created new lifestyles and new types of militant activity in alternative communities. They have now taken refuge in Friedrichshain and, although the area around Kottbusser Tor continues to look like a little Istanbul, the western part of the district is particularly delightful for its ancient façades clustered around Viktoriapark.

AL KALIF

GOLGATHA

RESTAURANTS

Curry 36 (D B3)
→ *Mehringdamm 36*
Mon–Fri 9am–4am; Sat 10am–4am; Sun 11am–4am
A grilled sausage in a *Brötchen*, with onions (*Zwiebeln*), smothered in ketchup and curry sauce – this is Berlin's astonishing specialty, to eat on the go. Around €2.

Al Kalif (D C4)
→ *Bergmannstr. 105*
Tel. 694 47 34
Daily noon–midnight
Take a relaxing break in Syria: remove your shoes when you go in and stretch out on the rugs and cushions. Excellent homemade dishes prepared using fresh produce. Couscous, hummus and falafel, served with mocha café or cinnamon tea – no alcohol. Dishes €5–20.

Osteria no. 1 (D A4)
→ *Kreuzbergstr. 71*
Tel. 786 91 62
Daily noon–midnight
Kreuzbergstrasse is Berlin's Little Italy and Osteria no. 1 a warm, classy, good-value Italian restaurant, which has swiftly become extremely popular. Generous portions and attractively presented dishes – *vitello tonnato* (minced veal in tuna and caper sauce), pasta, pizza; also, at last, good coffee. Reservation advised. Dishes €8–16.

Amrit (D F2)
→ *Oranienstr. 202-203*
Tel. 612 55 50 Daily noon– midnight (1am Fri-Sat)
This vast space is a café, restaurant and cocktail bar all at once. The menu features a wide choice of good, spicy Indian dishes. The tandoori duck is especially recommended. Dishes €8–14.50.

Austria (D C4)
→ *Bergmannstr. 30*
Tel. 694 44 40
Daily 6pm–midnight
A traditional Austrian *Gaststätte*, renowned for the quality of its cooking: *Wienerschnitzel* and *Strudel* of course, but also many other specialties. Reservation necessary. Dishes €9–17.

Altes Zollhaus (D D3)
→ *Carl-Herz-Ufer 30*
Tel. 692 33 00
Tue-Sat 6–11pm
One of the best restaurants in Berlin, in a half-timbered tavern on the banks of the canal. Light, innovative German cuisine prepared with the finest produce. The Brandenburg duck *magret* with mashed

ANKERKLAUSE   LE BATEAU IVRE   FASTER PUSSYCAT

potato and cabbage is delicious. Superb wine list, ranging from ordinary pitchers to bottles of Romanée-Conti. Menu €35–50.

BIERGARTEN, CAFÉS

Café Adler (D B1)
→ Friedrichstr. 206
Tel 251 89 65 Daily 10am–midnight (7pm Sun)
Overlooking the old Checkpoint Charlie (hence its fame and popularity) is the small, atmospheric, chic and slightly faded Café Adler. It serves German and continental cooking, but you can go for just a coffee or a beer. It is easy to see why John Le Carré (Tom Clancy too, apparently) stopped by so often. Menu €18.

Golgatha (D A4)
→ Katzbachstr., Viktoriapark (opposite from the football pitch)
Tel 785 24 53
April-Sep: daily 10am–6am
The Golgatha Biergarten, at the foot of the Kreuzberg, is one of the most famous of the 'beer gardens' that flourish in summer on sidewalks and in parks. Every night from 10pm, it becomes a club or a concert venue.

Locus (D C4)
→ Marheinekeplatz 4

Tel. 69 156 37
Daily 10am–1.30am
Enjoy a hearty Frühstück, away from the hustle and bustle of the market. With varied music in the background and a pretty terrace overlooking the square.

Ankerklause (D E3)
→ Kottbusser Damm 104
Tel. 693 56 49 Mon 4pm–5am; Tue-Sun 10am–5am
On the banks of the Landwehrkanal, this white and blue café, resembling a barge, is a pleasant spot in the afternoon. Later, it is crowded with people enjoying easy-listening tracks on the jukebox.

Le Bateau Ivre (D F2)
→ Oranienstr. 18
Tel. 61 40 36 59
Daily 9am–3am
Talking at the bar or debating under the lanterns, a cultured and fashionable clientele gather here, and the café is always crammed. You might even bump into a German movie star.

CLUBS, CONCERTS

SO 36 (D F2)
→ Oranienstr. 199
Tel. 61 40 13 06
Daily: call for info on concerts; www.so36.de
This concert hall and club

is a compulsory destination if you're touring the Berlin night scene. Eclectic program from hip-hop to ballroom dancing (Sun), and special lesbian or hard rock evenings.

Sage Club (D E1)
→ Köpenicker Str. 76
Tel. 278 98 30
Thu from 7pm, Fri-Sun from 11pm; www.sage-club.de
One of the swankiest nightclubs in town, in a former subway station. Black-suited doormen, VIP lounge, and a trendy clientele. Rock on Thu, electro on Sun. The club also hosts the wild evenings organized by the Kitkat Club (Fri-Sat).

Cake Club (D E2)
→ Oranienstr. 31
Tel. 616 59 399 Tue, Thu-Sun from 8pm (DJ from 10 pm); www.cakeclub.de
Funk, drum'n'bass, house, electro and delicious cocktails taken from the endless list (more than 100 cocktails) of the Cake Bar (Schlesische Str. 32, off map D F3).

SHOPPING

Faster Pussycat (D B4)
→ Mehringdamm 57
Tel. 69 50 66 00 Mon-Sat 11am–8pm (7pm Sat)

The best of stylish Berlin: retro but not backward. New-style wigs and retro coats vie for attention.

Colours (D B4)
→ Bergmannstr. 102 (rear of the courtyard, first floor on the left)
Tel 694 33 48 Mon-Sat 11am–7pm (6pm Sat)
This remarkable second-hand clothes supermarket sells everything: from classic jeans to wild psychedelic dresses, mini-skirts and T-shirts.

Markthalle Kreuzberg (D C4)
→ Bergmannstr., Zossener Str.
Mon-Fri 8am–7pm; Sat 8am–2pm
You can find everything in this tower of Babel: underwear and cigars, Turkish and Greek specialties, fresh produce and even animals! Built in 1822, the market hall is without doubt the meeting place of the area.

Kadò (D E3)
→ Graefestr. 75
Tel. 69 04 16 38
Tue-Fri 9.30am–6.30pm; Sat 9.30am–3.30pm
An old, quaint little grocer's store, where everything is made from liquorice. Sweet, salty or tangy, you inevitably leave with some.

SO 36 VIERTEL

↑ Map C

JÜDISCHES MUSEUM

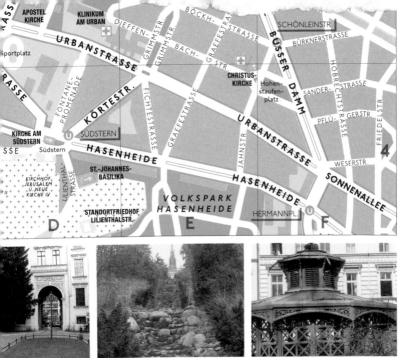

VIKTORIAPARK

CHAMISSOPLATZ

maze of spotless white rooms and corridors open into exhibitions of modern art. Each artist is given a studio for a year for the purpose of creating a work that will be exhibited after their stay.

SO 36 Viertel (D E2)
The former 'Süd-Ost 36' is the hub of working-class Kreuzberg. The traditional May 1 demonstrations express the political engagement of some of the district's inhabitants. The rest of the year, life revolves quietly around the markets (Kottbusser Tor and Görlitzer Park), the *Imbiss*, Turkish grocery stores (Oranienstr. and Kottbusser Damm) and

along the Landwehrkanal.
★ Schwules Museum (D B4)
→ *Mehringdamm 61*
Tel. 693 11 72
Wed-Mon 2–6pm (7pm Sat)
Before the Nazis, Berlin was home to the largest gay community in Europe. After the war, although in West Germany *Schwule* and *Lesben* regrouped in Schöneberg, East Berlin remained the only city in East Germany where they were tolerated. The Gay Museum opened its doors in 1988: the first floor is now dedicated to the permanent exhibition, the ground floor to the temporary ones; there's also a richly stocked library.

★ Viktoriapark (D A4)
This neo-Romantic park laid out on the highest hill in Berlin (1888–94) is topped by the National Monument, which commemorates the war of liberation against Napoleon. The Gothic spire built by Schinkel (1817–21) is graced by an iron cross (*Kreuz*) from which the district took its name.
★ Chamissoplatz (D B4)
→ *Organic market here on Sat 8am–2pm*
Bordered by *Mietskasernen* with delicately hued neo-Renaissance façades, and dotted with well-preserved turn-of-the-century urinals – a miracle in this badly bombed area, this is the

most typical square of pre-war Berlin.
★ Riehmers Hofgarten (D B3)
→ *Between Grossbeerenstr., Yorckstr. & Hagelberger Str.*
Neo-Renaissance façades facing wide tree-lined avenues: this is how 19th-century Berlin should have looked. The 328-foot-long plots of land provided by James Hobrecht in his plan (1856) were to have been reached by small inner streets. To cut costs, the developers crowded the cramped rental apartment blocks (*Mietskasernen*) around small dark courtyards. Only Wilhelm Riehmer did differently.

GEMÄLDEGALERIE

NEUE NATIONALGALERIE

★ **Hamburger Bahnhof Museum für Gegenwart Berlin** (E D1)
→ *Invalidenstr. 50-51*
Tel. 397 83 40
Tue-Fri 10am–6pm; Sat-Sun 11am–6pm (8pm Sat)
This museum of modern art is in the city's oldest railway station (1847). The metal structure of the vast hall and the side wings, renovated by Kleihues (1984), provide a fitting arena for installations by Beuys and artworks by Warhol.

★ **Reichstag** (E D2)
→ *Platz der Republik 1*
Tel. 22 70 Mon-Fri 9am–5pm (6pm in summer); Sat-Sun and public hols 10am–4pm (9am–6pm in summer); dome: daily 8am–10pm
Built in 1894, the building witnessed the death throes of the Weimar Republic before being torched by the Nazis on Feb 27, 1933. A symbol of the fall of Berlin, it was not until reunification that it welcomed the Bundestag under its new glass cupola (Norman Foster, 1999). Solemn and imposing, it attracts many strollers who've come to enjoy its esplanade.

★ **Potsdamer Platz** (E D4)
The biggest crossroads in Europe and the heart of the residential city in the 1920s, this square was razed to the ground by bombs. In the past ten years, this vast wasteland, bisected by the Wall, has become a flourishing business center.

Sony Center
→ *www.sonycenter.de*
The headquarters of Sony Europe, designed by Helmut Jahn. Beneath the vast tent over the buildings, light plays with the water features, next to the neo-Baroque vestiges of the Grandhotel Esplanade, which was put onto jacks and air cushions and moved aside so that the Center could be built.

★ **Gemäldegalerie** (E C4)
→ *Kulturforum, Matthaikirchplatz 4*
Tel. 266 29 51 Tue-Sun 10am–6pm (10pm Thu)
In the naturally lit rooms of the Kulturforum (1998), Caravaggio follows Giotto and Cranach follows Van Eyck for a comprehensive overview of 13th- to 18th-century European painting. Sadly, some 400 canvases from the Hohenzollerns' huge collection were lost during the bombing of the Friedrichshain bunker. Next door, there is a fine museum of decorative arts from the Middle Ages to the present day.

★ **Neue Nationalgalerie** (E C4)
→ *Kulturforum, Potsdamer S Tel. 266 2951 Tue-Sun 10am–6pm (10pm Thu-Sat; 8pm Su* Picasso, Ernst, De Chirico. This glasshouse for moder

E

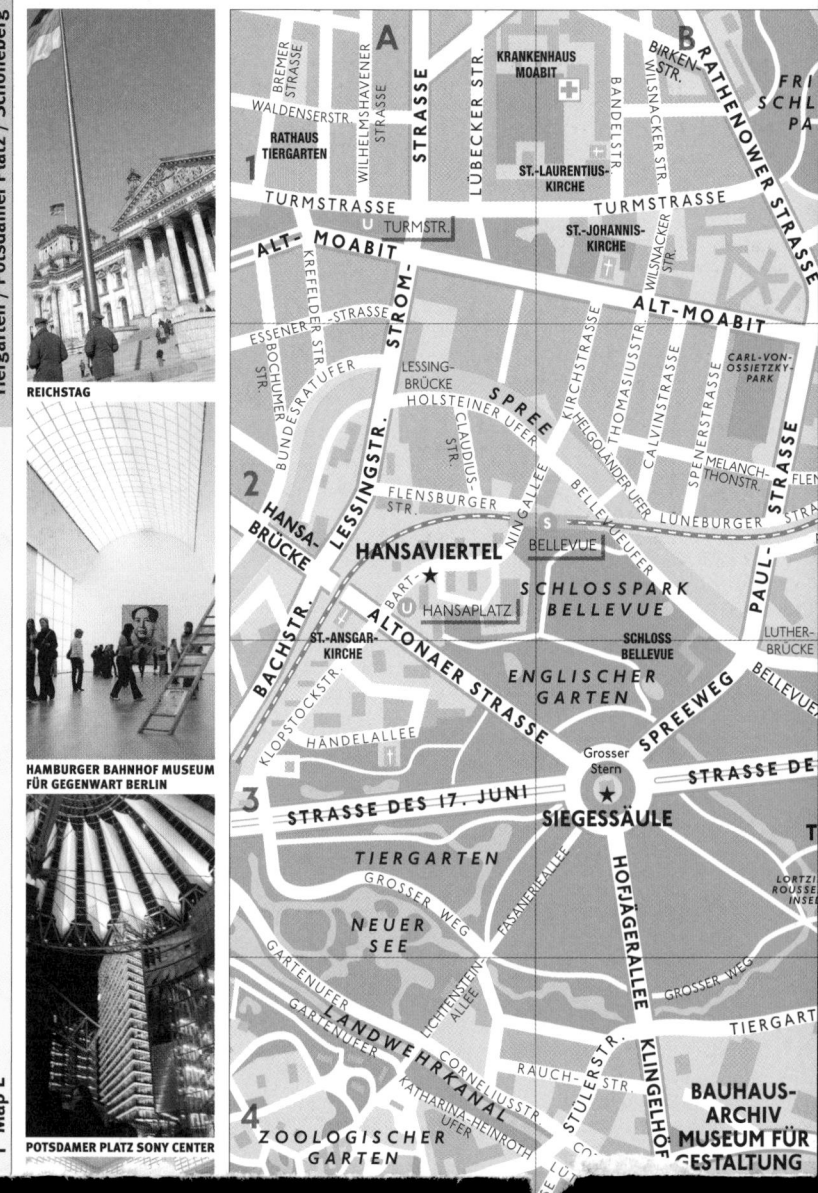

REICHSTAG

HAMBURGER BAHNHOF MUSEUM
FÜR GEGENWART BERLIN

POTSDAMER PLATZ SONY CENTER

Surrounding the Siegessäule and the Schloss Bellevue, now the president's home, the Tiergarten, former hunting preserve of the Hohenzollerns, is a cool oasis of greenery in the city and a popular destination for summer barbecues. Further south, the buildings on the new Potsdamer Platz form an extension of the Kulturforum and serve as a monument to leading contemporary architects. By comparison, the peaceful Schöneberg district appears extremely unassuming. However, appearances can be deceptive: the gay community parties into the early hours of the morning in the many bars and clubs.

PAN Y TULIPAN

KUMPELNEST 3000

RESTAURANTS

Goltzstrasse (E B6)
Two addresses among many on Schöneberg's restaurant-lined street.
Shayan
→ *Goltzstr. 23; Tel. 215 15 47*
Daily noon–midnight
A welcoming family atmosphere and very good Indo-Iranian cuisine. Dishes €4.50–15.
Indischer Imbiss
→ *Goltzstr. 33*
Tel. 215 49 65 Daily 11am–1am (2am Fri-Sat)
This restaurant's kitsch decor must be seen: a Hindu altar, incense, glitzy statuettes, hangings and garlands of lights. Healthy food and generous portions. Dishes €5–9.
Tiergartenquelle (E A3)
→ *Bachstr. 6 (under the Tiergarten subway station)*
Tel. 392 76 15
Mon-Fri 5pm–midnight; Sat-Sun noon–midnight
Near Hansaplatz, this friendly, family-run *Kneipe* is located under the arches of the S-Bahnhof Tiergarten. Traditional German dishes: platter of cooked meats, Nuremberg sauerkraut or escalopes. Dishes €6–12.
Pan y Tulipan (E B6)
→ *Winterfeldstr. 40*
Tel. 21 91 30 14 Daily 10am

(9am Sat)–midnight
A very warm and friendly service – you can't help but smile back, and Spanish cuisine in generous portions. There's a lounge on the first floor for more intimate meals on snug sofas. Dishes €10–16; a few tapas at cheaper prices.
Lutter & Wegner im Kaisersaal (E D4)
→ *Potsdamer Platz*
Tel. 26 39 03 82
Daily noon–midnight
The best of Austrian and German cuisine in a hip, suave decor in the heart of busy Potsdamer Platz. Dishes from €17.
Café Einstein (E B5)
→ *Kufürstenstr. 58*
Tel. 261 50 96
Daily 9am–midnight
A faultless classic: Austrian cuisine in a Jugendstil villa. In summer *Wiener Schnitzel* and Viennese coffees are served in the garden.
À la carte € 30.
Lochner (E B4)
→ *Lützowplatz 5*
Tel. 23 00 52 20
Tue-Sun 6–10.30pm
As stylish as it is spacious, this up-market restaurant has sleek marble flooring and long, wooden tables. The menu offers a creative blend of German and Mediterranean flavors.

PHILHARMONIE

DECO ARTS

HARB

Set menu €45.

Daitokai (E A5)
→ *Tauentzienstr. 9-12*
Tel. 261 80 90
Daily noon–2pm, 6–10pm
Excellent Japanese cuisine and impeccable service. Menu €49.

CAFÉS, BARS

Joseph Roth Diele (E C5)
→ *Potsdamer Str. 75*
Tel. 26 36 98 84
Mon-Fri 10am–midnight
Office workers, students and passerby come to this old taverna for the famous *Gulashsuppe*. And there's always someone to play a tune on the piano.

Walhalla (E A1)
→ *Krefelder Str. 6 (corner of Essenerstr.); Tel. 393 3039*
Daily 10am–2am;
www.walhalla-berlin.de
This *Kneipe* is always full: it has a pool table, a terrace looking onto the street when the weather is good, some good dishes and fine beers, i.e. all the right things to make it a favorite among locals.

Käfer's-Dachgarten Restaurant (E D2)
→ *Platz der Republik*
Tel. 22 62 99 33
Daily 9am–midnight
Modern-looking space on the roof of the Reichstag and fantastic views over

the city. The German-continental cuisine is average so go for breakfast. Reservation is essential and will allow you to jump the queue to see the dome (entry via the West portal).

Qiu (E D4)
→ *Potsdamer Str. 3*
Tel. 590 05 12 30 Daily noon–2am (1am Sun)
Situated on the first floor of the Mandala Hotel, this lounge dominates Potsdamer Platz. Great cocktails.

Café M (E B6)
→ *Goltzstr. 33*
Tel. 216 7092 Mon-Fri 8am–2am; Sat-Sun 9am–3am
Lively mixed bar (straight and gay), with a background of electronic and jungle music. Drink White Russians, Mojitos or beers.

CONCERT HALL, THEATER, CLUB

Philharmonie (E D4)
→ *Herbert-von-Karajan-Str. 1*
Tel. 25 4880 (general info)
Ticket office: Mon-Fri 3–6pm; Sat-Sun 11am–2pm; www. berlin-philharmonic.com
A marriage made in heaven: the huge amphitheater-shaped auditorium, designed by Hans Scharoun, offers the

illustrious Berliner Philharmonisches one of the finest acoustics in the world. Next door, chamber music at the Kammermusiksaal.

Grips Theater (E A2)
→ *Altonaer Str. 22 (Hansaplatz); Tel. 397 47 40*
Ticket office: Mon-Fri noon–6pm; Sat-Sun 11am–5pm; www.grips-theater.de
People of all ages pack into this amphitheater-style auditorium. Plays by the Grips company – politically committed yet good-natured – provide musical thumbnail sketches of Berlin life, such as the legendary *Line 1*.

Kumpelnest 3000 (E C5)
→ *Lützowstr. 23*
Tel. 261 69 18 Daily 7pm–5am (7am Fri-Sat)
Ultra-kitsch decor and music: the barman is an expert on the Deutsche Schlager – schmaltzy German songs from the 1970s – and he also knows the 50 most recent winners of the Eurovision Song Contest by heart. Laid-back and friendly atmosphere.

SHOPPING

Garage (E B5)
→ *Ahornstr. 2; Tel. 211 27 60*
Mon-Fri 11am–7pm (8pm

Thu-Fri); Sat 11am–6pm
This store has the cheapest second-hand clothes in the city, but not always the latest fashions. Everything is sold by weight; 'happy hour' Wed 11am–1pm.

Deco Arts (E B6)
→ *Motzstr. 6*
Tel. 215 86 72 Wed-Fri 3–7pm; Sat noon–5pm
This store specializes in 20th- and 21st-century furniture and decoration: Verner Panton lamps, Breuyers armchairs, Scandinavian furniture from the 1950s–60s, English sofas, as well as pretty vases, accessories and unusual articles.

Antike Möbel (E B6)
→ *Goltzstr. 49*
Tel. 216 37 19 Mon-Fri 1–7pm; Sat 10am–3pm
This antiques dealer and restorer has some fine pieces of German antique furniture. He also sells beeswax candles made using traditional methods.

Harb (E C5)
→ *Potsdamer Str. 93*
Tel. 261 19 36 Mon-Sat 9am–7pm (5pm Sat)
An Ali Baba's cave full of exquisite crockery, lamps, music, incense and teas. For the gourmets: fresh olives and dried fruits, marinaded vegetables and Oriental patisserie.

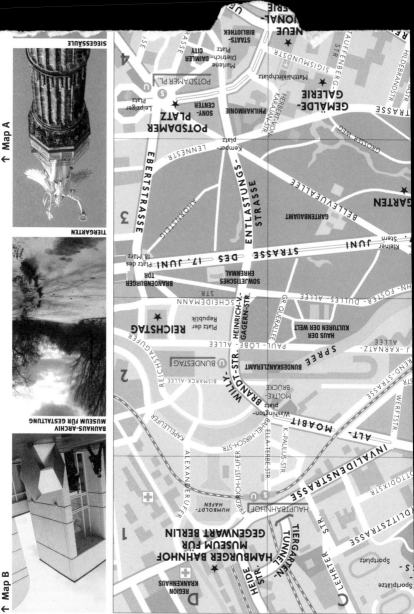

HANSAVIERTEL

SCHÖNEBERG

→ **Map D**

art, built by Mies van der Rohe in 1968, has a large Expressionist section: Kokoschka and Munch, and German painters from Die Brücke (Kirchner, Nolde) and Die Neue Sachlichkeit ('new reality') – Dix, Grosz and Beckmann.

★ **Bauhaus Archiv-Museum für Gestaltung (E** B4)
→ *Klingelhöferstr. 14*
Tel. 254 00 20
Wed-Mon 10am–5pm
Functionalism, rejection of ornamentation and geometrization: the building constructed in 1979 to plans by Walter Gropius applies principles advocated by the Bauhaus, which Gropius ran from 1919 to 1932. The

museum exhibits paintings and models by artists from this school, which united art, design and architecture: Kandinsky, Breuyer, etc.
★ **Tiergarten (E** C3)
This green lung of 412 acres, in the heart of the city, is the old hunting reserve of the Hohenzollerns. It transforms into a huge barbecue in summer. The park houses the Schloss Bellevue and the House of World Culture, architectural testament to the 1950s.
★ **Siegessäule (E** B3)
→ *Strasse des 17. Juni*
Nov-March: daily 10am–5pm;
(5.30pm Sat-Sun);
April-Oct: daily 9.30am–
6.30pm (7pm Sat-Sun)

From the top of its 194-foot-high pedestal decorated with cannons taken from Prussia's enemies, 'Goldelse' watches over the Tiergarten. The golden angel from Wim Wenders' *Wings of Desire* commemorates Prussia's victories over Denmark (1864), Austria (1866) and France (1871). Originally set up opposite the Reichstag, it was moved here by Hitler in 1938. There are breathtaking panoramic views from the top.
★ **Hansaviertel (E** A2)
→ *Map of the project on Hansaplatz*
In 1953, as a counterpart to the building sites on Karl-Marx-Allee, West Berlin

commissioned 48 architects (Gropius, Niemeyer, Aalto...) to build 1,400 apartments. These apartment blocks, scattered amidst greenery around a store-lined square, marked the apex of the town-planning projects developed in Berlin as a reaction to the *Mietskasernen*.
★ **Schöneberg (E** B6)
The ageing façades and modern buildings in this charming little residential district are home to one of the largest gay communities in Europe. In the trendy bars on Goltzstrasse and the surrounding area, no one bats an eyelid at the prominent number of gay and lesbian couples.

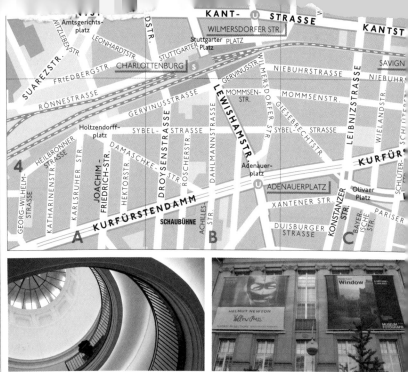

SAMMLUNG BERGGRUEN

MUSEUM FÜR FOTOGRAFIE

★ Kaiser-Wilhelm-Gedächtniskirche (F E3)
→ Breitscheidplatz
Tel 21 47 63 21
Daily 9am–7pm
Since 1943, this 'Memorial church' (1890–5) has been pointing its damaged steeple skyward, reminding passersby of the wreckage caused by bombing (over a third of the city's houses were destroyed). At night, the tower and nave of the new church (1961) form two prisms of blue glass illuminated from within.

★ Zoologischer Garten (F F3)
→ Budapester Str. 34
Tel 25 40 10 Daily
9am–6.30pm (5pm winter)
Bison wander around an Indian totem pole, the giraffe house is reminiscent of a mosque and the mouflon enclosure resembles an alpine crest. Behind its pagoda-like entrance gate the Lenné Zoo (1844), with 15,000 animals (including two pandas), is one of the world's largest.

★ Käthe-Kollwitz-Museum (F D4)
→ Fasanenstr. 24
Tel. 882 52 10
Wed–Mon 11am–6pm;
www.kaethe-kollwitz.de
The Expressionist work of Käthe Kollwitz (1867–1945) conveys her horror at the poverty she discovered with her husband, a doctor, in Prenzlauer Berg. As most of her sculptures were destroyed during World War Two, the museum largely exhibits her engravings.

★ Savignyplatz (F D3)
A 'forest' of parasols surrounds large, peaceful, tree-lined streets with Jugendstil façades. Late into the night, lively cafés are packed with students from the nearby universities.

★ Bröhan-Museum (F A1)
→ Schloss-Str. 1a
Tel. 32 69 06 00
Tue–Sun 10am–6pm;
www.broehan-museum.de
Gallé vases, porcelain from Copenhagen, furniture by Guimard, Van de Velde or Ruhlmann. Karl H. Bröhan's collection, devoted to designs from 1889 to 1939, features the virtuoso arabesques of Jugendstil, and the spare lines of Art Deco.

★ Sammlung Berggruen (F A1)
→ Schloss-Str. 1
Tel. 326 95 80
Tue–Sun 10am–6pm;
www.smb.spk-berlin.de
The greatest names in modern art collected by Heinz Berggruen, a shrewd gallery owner: Cézanne, Matisse, Giacometti, Klee. Over 80 works spanning Picasso's entire career, from the Blue period to the work in ink of 1971, include a study for the Demoiselles

F

KÄTHE-KOLLWITZ-MUSEUM

ZOOLOGISCHER GARTEN

KAISER-WILHELM-GEDÄCHTNISKIRCHE

The 'Ku'Damm', at the center of West Berlin, is like a capitalist Unter den Linden. Today, where cabarets and cafés once drew a bohemian crowd in the 1920s, a string of rather characterless buildings houses a variety of stores owned by large international fashion brands. However, even since the fall of the Wall, droves of tourists and Berliners continue to take the district's main thoroughfares to the shops and theaters. Further west, the middle-class residences of Charlottenburg – small apartment blocks and luxury villas – cluster around the Baroque castle and its adjoining museums.

DAO THAÏ

CAFE IM LITERATURHAUS

RESTAURANTS

Dao Thaï (F C3)
→ Kantstr. 133
Tel. 37 59 14 14
Daily noon–11pm
Casseroles simmer behind the counter: Thai specialties served directly from the pan to your plate. Dishes €5.90–15.

Good Friends (F C3)
→ Kantstr. 30
Tel. 313 26 59
Daily noon–2am
The waiters are in a constant flurry as they glide through the packed dining room. More than 100 dishes are on offer at this hangout of Berlin's Chinese community. Dishes €7–14.

Engelbecken (F A3)
→ Witzlebenstr. 31
Tel. 615 28 10
Mon–Fri 5– 11pm; Sat 4–11pm; Sun noon–11.30pm
A quality restaurant with a menu based on Bavarian and Alpine specialties including schnitzel and scrumptious beef and lamb stews. The ingredients are organic and ethically sourced. Try to reserve an outside table, weather permitting. Dishes €8–18.

Scarabeo (F D4)
→ Ludwigkirchstr. 6
Tel. 885 06 16   Daily
4pm–1am (2am Fri-Sat)
A rather kitsch decor, belly dancing, but mainly succulent Egyptian specialties. The non-gourmets come to smoke water pipes under the wise eye of Nefertiti. As a starter, do try the mezze (€22); couscous from €10.

Trattoria
à Muntagnola (F F4)
→ Fuggerstr. 27
Tel. 211 66 42
Daily 5pm–midnight
The Muntagnola ('small mountain') is a perfect name for this place. With grandma's crockery, images of the Madonna and the patron's funny jokes, this Italian trattoria does have a rustic charm. It has links with a group of children with AIDS, who are invited once a week to have a tasty bowl of pasta. Dishes €10–21.

Cafe im
Literaturhaus (F D4)
→ Fasanenstr. 23
Tel 882 54 14
Daily 9.30am–1am
Fine gravel paths winding through lawns and banks of flowers, a Jugendstil glasshouse made of wrought iron, and staff dressed in white suits with bow-ties: take a break from the frenetic activity of the Ku'Damm to recall what it

QUASIMODO STILWERK KU'DAMM

was like when people took their time. Global cuisine and a good wine list. Dishes €15–25.

CAFÉS, BARS, TEAROOMS

Kleine Orangerie (F A1)
→ Schloss Charlottenburg
Tel. 322 20 21 Daily
9am–8pm (6pm winter)
Opposite the museums, at the far end of the castle's orangery, are an exotic glasshouse and a classic *Kneipe* for a coffee break, a *Flamküche* or a *Frühstücksbüfett* (served Sun 10am–2pm).

Gainsbourg (F D3)
→ Savignyplatz 5
Tel. 313 74 64
Daily from 4pm
Situated in one of Berlin's most pleasant squares, this atmospheric bar pays homage to the French singer Serge Gainsbourg. Close by Savignyplatz 1, and on a similar theme, is Café Brel.

Zwiebelfisch (F D3)
→ Savignyplatz 7-8
Tel. 312 73 63
Daily noon–6am
A cult *Kneipe* since May, 1968, with an old clientele recalling their exploits while reading the newspapers, often arguing about politics in

the smoky atmosphere.

Wirtshaus Wubke (F C3)
→ Schlüterstr. 21
Tel 31 50 92 17
Daily 11am–3am
A local *Kneipe* frequented by students, artists and regulars who will gladly strike up a conversation at the bar over a schnapps '43' or a draft beer.

BARS, CLUBS

Die Kleine Weltlaterne
(off map F B4)
→ Nestorstr. 22
Tel. 89 09 36 16
Mon-Sat 8pm–3am
A Berlin myth since 1961! This late bar has acquired a reputation for spotting young talent, including painters and musicians. Even if it doesn't have the crowds of its heyday, the customers are loyal and a good ambience is assured. Jazz concerts (Thu and Sat).

A-Trane (F C3)
→ Pestalozzistr. 105
Tel. 313 25 50
Daily 9pm–2am (4am Fri-Sat); concerts: daily 10pm
Small concert hall for a big club that bills the top jazz performers, but is keen to promote local talent.

Quasimodo (F D3)
→ Kantstr. 12a
Tel. 312 80 86 Tue-Sat 9pm;

concerts 10pm (advance ticket sales from 11am); bar open daily from 11am; www.quasimodo.de
Under the café is the jazz club, which not only puts on stars but also holds legendary jam sessions: jazz, soul, disco, blues. The audience is young(ish) and the mood is charged.

OPERA HOUSE

Deutsche Oper (F C2)
→ Bismarckstr. 35
Tel. 343 84 01
Ticket office: Mon-Sat 11am–one hour before the show; Sun 10am–2pm; www.deutscheoperberlin.de
West Berlin's opera house: lavish productions and international stars.

SHOPPING

Ku'Damm (F D4)
→ Kurfürstendamm
The commercial hub of Charlottenburg: more than 2 miles of stores, art galleries, luxury hotels and theaters, which attract people of all ages.

New Steinbruch (F E4)
→ Kurfürstendamm 237
Tel. 88 55 21 26
Mon-Sat 10am–8pm
Satin pants, T-shirts and high boots for drag queens.

Stilwerk (F C3)
→ Kantstr. 17
Tel. 31 51 50 Mon-Sat 10am–8pm (6pm Sat)
A commercial design center on five floors, all dedicated to homeware: big brands, but also local designers, antiquarians and even a piano supplier. Pretty interior garden.

Crines design (F D4)
→ Kurfürstendamm 39
Tel. 88 77 48 44
Mon-Sat 10am–8pm
A Berlin designer with a discreet, classic and elegant style, using linen in all its forms, and raw or woven material that she makes into expansive tunics and large hats.

Ka De We (F F4)
→ Tauentzienstr. 21-24
Tel. 212 10
Mon-Fri 10am–8pm (10pm Fri); Sat 9.30am–8pm; www.kadewe.de
This is Europe's largest department store (1907), with everything from haute couture to the complete range of Ritter Sport (chocolate bars in exotic flavors). The food hall on the top floor is a gourmet's paradise, with over a thousand types of ham and sausages, cheeses, breads and a buffet restaurant with panoramic views.

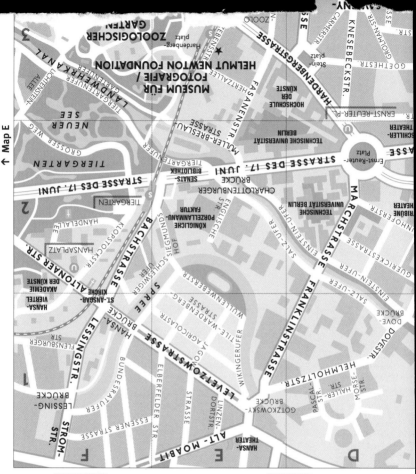

SAVIGNYPLATZ

BRÖHAN-MUSEUM

SCHLOSS CHARLOTTENBURG

MUSEUM FÜR VOR- UND FRÜHGESCHICHTE

SCHLOSSGARTEN

d'Avignon, cubist paintings and portraits, etc.

★ Museum für Fotografie / Helmut Newton Foundation (F E3)
→ *Zebensstr. 2*
Tel. 266 2188 Tue-Sun 10am–6pm (10pm Thu);
www.smb.spk-berlin.de
This 1909 neoclassical building successively housed the Landwehr Casino, the Beaux-Arts Library and, since June 2004, the Helmut Newton Foundation. Under the same roof as the foundation is the Photography Museum, showing temporary and permanent exhibitions of Newton's work but also a documentation and research center dedicated to modern photography, complete with library, restoration workshop and photographic laboratory.

★ Schloss Charlottenburg (F A1)
→ *Tel. 32 09 11*
Tue-Sun 9am–5pm
Berlin's Baroque gem (1695–1701), dedicated by Frederick I (1688–1713) to his wife Sophie Charlotte and enlarged in 1740–6 by Knobelsdorff for Frederick II. Inside are the king's sumptuous apartments, his French painting collection (including eight Watteaus), the Chinese and Japanese porcelain room and, in the Galerie der Romantik, masterpieces by Romantic German painters such as Caspar D. Friedrich.

★ Museum für Vor- und Frühgeschichte (F A1)
→ *Schloss Charlottenburg*
Tel. 32 67 48 40 Tue-Fri 9am–5pm; Sat-Sun 10am–5pm
From the Paleolithic Period to the Germanic Early Middle Ages, the complete pre- and protohistory of Europe and the Near East. Caucasian ceramics, a ceremonial gold helmet from the Bronze Age and copies of the finest pieces from the treasure of Troy (jewelry and gold crockery) discovered by Schliemann, and seized by the Russians in 1945.

★ Schlossgarten (F A1)
→ *Tel. 32 09 14 40*
Belvedere: April-Oct: Tue-Fri noon–5pm; Sat-Sun 10am–5pm; Nov-March: Tue-Sun noon–4pm; Mausoleum: April-Oct: Tue-Sun 10am–5pm
Having strolled through the Baroque garden by Siméon Godeau (a pupil of André Le Nôtre), you can explore the paths and footbridges of the landscaped garden (1819–1928), designed by Lenné. Don't miss the Neapolitan pavilion by Schinkel, the Belvedere by Carl Gotthard Langhans and the mausoleum of Queen Louise (1776–1810), the mother of Prussian nationalism.

FLYING TO BERLIN

From the US
Lufthansa
→ Tel. 800 645 3880
www.lufthansa.com
Delta Airlines
→ Tel. 800 221 1212
www.delta.com
Also check the following:
www.cheapflights.com
From the UK
Lufthansa
→ Tel. 0845 7737 747
www.lufthansa.co.uk
AirBerlin
→ Tel. 0870 738 8880
www.airberlin.de
Ryanair
www.ryanair.com
AirBerlin and Ryanair fly
from London Stansted to
Berlin Schönefeld.

TRAIN AT THE ZOOLOGISCHER GARTEN STATION

TAXI RANK ON SAVIGNYPLATZ

€80–110

Arco Hotel (E A6)
→ Geisbergstr. 30,
Schöneberg; Tel. 235 14 80
www.arco-hotel.de
A small hotel in a peaceful
street. Comfortable rooms.
Breakfast in the garden
in summer. €77–95.
Hotel Gunia (E B6)
→ Eisenacherstr. 10,
Schöneberg; Tel. 218 59 40
www.hotelgunia.de
High ceilings, moldings
and gilt: this 300-year-old
building, listed as a historic
monument, has gorgeous
rooms full of character.
€75–100.
East-Side Hotel (C D4)
→ Mühlenstr. 6,
Friedrichshain
Tel. 29 38 33
www.eastsidehotel.de
Behind a listed façade,
this hotel-cum-art gallery
exhibits paintings in the
hall and dining room. There
are no paintings in the 36

modern rooms, but those
overlooking the street have
a view of... the East Side
Gallery. From €85.
**Hotel am
Scheunenviertel (B** B4)
→ Oranienburgerstr. 38,
Mitte; Tel. 282 21 25
www.hotelas.com
A well-situated if rather
conventional hotel: perfect
for night revelers as it is
right in the heart of Mitte.
If peace and quiet is
paramount, ask for a room
overlooking the courtyard.
From €91.
**Artist Riverside
Hotel & SPA (A** C1)
→ Friedrichstr. 106, Mitte
Tel. 284 900
www.tolles-hotel.de
Near Oranienburgerstrasse
and Unter den Linden, this
Jugendstil hotel has been
decorated to resemble a
theater or movie theater.
Some rooms offer splendid
views over the Spree.
Discounts for artists and

musicians. Spa. €90–130;
breakfast €9.
**Hotel-Pension
Kastanien Hof (B** D3)
→ Kastanien Allee 65,
Prenzlauer Berg; Tel. 44 30 50
www.kastanienhof.biz
A well-located hotel, set
between Mitte and
Prenzlauer Berg; it is also
very near the delightful
Zionskirchplatz. Large,
bright rooms from €103.

€110 and above

MitART Hotel (B B4)
→ Linienstr. 139, Mitte
Tel. 28 39 04 30
www.mitart.de
A hotel of the 21st century,
the MitART aims to be
environmentally friendly
where possible (natural raw
materials, organic food,
natural cleaning agents
etc.), without compromising
comfort. It is quiet, close
to the city center and the
30 rooms spread over

TRAINS

Stations of arrival
→ Tel. 118 61
Zoologischer
Garten (F E3)
→ Hardenbergplatz 11
To the west (near the zoo).
Ostbahnhof (C C3)
→ Friedrichshain
To the east.
Hauptbahnhof (E D1)
The city's central station,
north of the Tiergarten
district, opened in 2006.

CARS

Wide, modern streets,
designed with cars in
mind. Avoid roadworks
(frequent) and rush hours
(7–9am, 4–6pm).
Driving
The Berliners always
abide by the highway
code: stop at orange
lights, give way to cyclists
and pedestrians on
crosswalks. At crossroads,
the first vehicle there has
right of way.
Speed limits
20 to 30 mph in the city.
No speed limit on
freeways.
Parking
→ €1–3/hr; €20/day
It is difficult to park in the
center. Illegally parked
cars will be towed away.

TAXIS

Fares
→ Pick up charge €3, then
€0.10/km
Short journey
→ Fixed price €3
(max. 2 kms/1⅕ miles)
Reservations
Würfelfunk Tel. 210 101
Taxi Funk Tel. 443 322
Funk Taxi Tel. 261 026
Cityfunk Tel. 210 202

AIRPORTS

→ www.berlin-airport.de
Tegel (TXL)
International flights
→ By bus 109, 128, X9 or
TXL; approx. 40 mins; €2.10
→ By taxi: 20 mins; €15–20
Tempelhof (THF)
Domestic flights
→ By bus 248 to
Alexanderplatz; €2.10
→ By U-Bahn line 6;
15 mins; €2.10
→ By taxi: 15 mins; €20
Schönefeld (SXF)
International flights
→ By AirportExpress train to
Ku'damm in 30 mins; €2.40
→ By S-Bahn (S45 or S9
lines)
→ By taxi: min. 40 mins;
€30–40

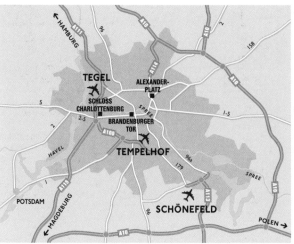

AIRPORTS AND MAIN HIGHWAY ROUTES

Unless otherwise stated, the prices given here are for a double room with bathroom and include breakfast. Reservations are essential six to eight weeks in advance from June to Sep, between Christmas and New Year's Eve, at Easter and when the city is hosting special events. Small pensions do not always accept credit cards.

PRIVATE ACCOMMODATION

Bed & Breakfast Berlin (D C2)
→ Feilnerstr. 1, Kreuzberg
Tel. 78 91 39 71
Mon–Fri 10am–6pm;
www.bed-and-breakfast.de
Apartment rentals and bed & breakfast options throughout the city (over 4,000 beds). Good facilities at modest prices and an excellent way of meeting Berliners. Rooms from €41; apartments from €54.

€50 and less

Youth hostels in Berlin
→ www.hostel.de
Globetrotter Hostel Odyssee (C E3)
→ Grünbergerstr. 23,
Friedrichshain; Tel. 2900 0081
www.globetrotterhostel.de
Extravagantly decorated hotel that's run by former backpackers, two minutes from Simon-Dach-Str. Bar-restaurant. Dorm €14–15.50/person; double room €47–54; breakfast €3.
Hotel Pegasus (C D3)
→ Str. der Pariser Kommune 35, Friedrichshain
Tel. 297 73 60
www.pegasushostel.de
Between Karl-Marx Allee and East Side Gallery, this hotel offers rooms with one to ten beds and one flat for two people. Self-service kitchen, breakfast (€5) in the garden. Dorm €10/ person + €2.50 for sheets; double room €60–70.

Amstel House (E A1)
→ Waldenserstr. 31
Tel. 395 40 72
www.circus-berlin.de
A modern hostel in an Art Nouveau building, 15 minutes from Tegel Airport by subway. Dorm €13/person; double room with en-suite from €25/person; sheets €3. Buffet breakfast €3.50.
Hotel Transit (D A4)
→ Hagelberger Str. 53,
Kreuzberg; Tel. 789 04 70
In the heart of Kreuzberg, two floors of dormitories and rooms around the courtyard of a former Mietskaserne. Dorm €32; double room with en-suite shower €72.

€50–80

Pension Kreuzberg (D B3)
→ Grossbeerenstr. 64,
Kreuzberg; Tel. 251 13 62
www.pension-kreuzberg.de
A family-run pension in a Gründerzeit building (1870s). Fine period staircase, spacious rooms, attentive service and young clientele. €55 (without bathroom)–65 (with).
Pension Peters (F D3)
→ Kantstr. 146,
Charlottenburg; Tel. 312 22 78
www.pension-peters-berlin.de
This family-run hotel owned by a sociable couple offers 36 bright, gaily colored rooms. They also rent out apartments in Mitte (min. three-day stay). €73–80.
Pension Funk (F D4)
→ Fasanenstr. 69,
Tel. 882 71 93
www.hotel-pensionfunk.de
A building once owned by Asta Nielsen, a star of silent movies , this hotel dates to the end of the 19th century, and has spacious rooms with Jugendstil windows and over-the-top decor. It's ideally located in a beautiful tree-lined street near the Ku'Damm. €72–113.

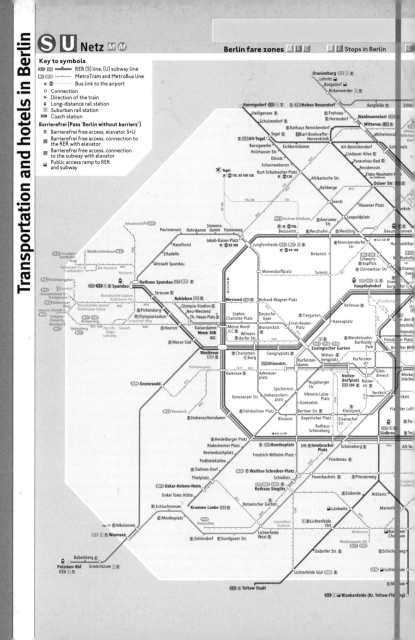

S U Netz M M

Berlin fare zones A B C   D E Stops in Berlin

Key to symbols

— RER (S) Line, (U) subway line
— MetroTram and MetroBus line
Bus link to the airport
○ Connection
► Direction of the train
Long-distance rail station
Suburban rail station
ZOB Coach station

Barrierefrei (Pass 'Berlin without barriers')

Barrierefrei free access, elevator S+U
Barrierefrei free access, connection to the RER with elevator
Barrierefrei free access, connection to the subway with elevator
Public access ramp to RER and subway

STREETCAR AT NIGHT

BUS

CYCLING IN FRONT OF THE REICHSTAG

Information
→ Tel. 194 49 (24/7)
The BVG runs the three areas of the network, except the S-Bahn, which is privately owned.
U-Bahn (subway)
Nine lines cover the city.
S-Bahn (commuter train)
15 train lines operate a service to the outskirts.
Bus
165 lines. Numerous double-decker buses.
Tram
27 modernized lines, traveling mostly in East Berlin.

When?
→ Daily 4am–2am
Nachtlinien (night lines)
→ Daily 2am–4am (every 30 mins)
Efficient night service: 56 bus lines, 5 tram lines and 2 U-Bahn lines.
Nightlinesurfer
Brochure listing the times of the Nachtlinien from the center, and the bars and clubs they pass.

Tickets
Ticket machines in U-Bahn stations and ticket offices in railroad stations.
→ www.visitberlin.de
Kurzstrecke
→ €1.20 for six stops (bus) or three subway stations (U-Bahn/S-Bahn)
Einzelfahrausweis
→ One-way ticket: €2.30, valid for two hours
Tageskarte
→ €5.70 for one day
7-Tage-Karte
→ €25.40 for one week
WelcomeCard
→ €21 for 72 hrs
Family ticket for one adult with three children.

---

www.luise-berlin.com
Each of the 50 rooms in this hotel is a work in its own right, designed by an artist whose name it adopts. For instance, there is the 'dream' by Dieter Mammel: an immense dark wooden bed covering the entire floor, with a canvas on the wall by the painter. Bear in mind that some rooms are reserved several months in advance. €135–149; breakfast €9.50.

Mandala Hotel (E D4**)**
→ Potsdamer Str. 3
Tel. 590 05 00 00
www.themandala.de
Formerly the Madison Postdamer Platz, the Mandala remains a winner. White, light, airy, hip design hotel meant for relaxation and Zen-itude. The rooms are vast and comfortable, staff are friendly and the prices aren't that high. Excellent restaurant giving onto the garden, cocktail lounge on the first floor and spa rooms on the 11th floor with views over the whole city. From €135.

Hecker's Hotel (F D3**)**
→ Grolmanstr. 35, Charlottenburg; Tel. 88 900
www.heckers-hotel.com
Centrally located, the Hecker's is a modern building with a very appealing interior, with 69 stylishly furnished rooms and excellent service. Magnificent views from the bay windows of the fifth-floor breakfast room. Roof terrace in summer. €140–280; breakfast €15.

LUXURY HOTELS

Hotel Adlon Kempinski (A B3**)**
→ Unter den Linden 77, Mitte
Tel. 226 111 11
www.hotel-adlon.de
Opposite the Brandenburg Gate, this legendary palace, inaugurated by William II, used to welcome Charlie Chaplin, Rockefeller and Marlene Dietrich. Rebuilt in 1997, it was refurbished to a decadently lavish standard and has just celebrated its centenary. The suites have a view over Unter den Linden or over Brandenburg Gate. From €330.

Schlosshotel im Grünewald (off map)
→ Brahmsstr. 10, Wilmersdorf
Tel. 89 58 40
www.regenthotels.com
This palace, more a country castle than a city hotel, dates from the 1910s and, with its marble and precious wood fittings, is one of the most luxurious hotels in Berlin. Its amazing refurbishment, in 1992, was directed by Karl Lagerfeld in return for a permanent, magnificent suite. One drawback: it is 30 minutes away from the city center by taxi (no buses or trams). From €225; breakfast €23.

Helmholtz-pl.

HAUPTBAHNHOF Alexanderplatz

OSTBAHNHOF

SCHLOSS CHARLOTTENBURG

SIEGESSÄULE ■ **BRANDENBURGER TOR**
100 249

142

ZOOLOGISCHER GARTEN

Oranien-pl.

FUNKTURM

248
129

RAILWAY STATIONS AND MAIN BUS ROUTES

CYCLES

Cycle lanes
On the sidewalk of most main roads.
U-Bahn and S-Bahn
It will cost you €1.50 to take your bike with you, in the U-Bahn.
Rental
Fahrradstation (A C2)
→ *Dorotheenstr., 30*
(four other branches)
Tel. 0180 510 8000 Mon-Sat 10am–7pm (4pm Sat);
www.fahrradstation.com
€15/day, €50/week.
Call a Bike
→ *Tel. 0700 05225522*
www.callabike.de
Rent a bike at all major crossroads, with a simple phone call. Max €15/day.

three floors (of which two are for non-smokers) are spacious, and simply but well decorated. Café on the ground floor with wi-fi access. €110–180; breakfast €7.

Hotel Honigmond (B B3)
→ *Tieckstrs. 12, Mitte*
Tel. 284 45 50
www.honigmond.de
This cozy pension is several minutes from Oranienburger Tor and has 40 bright, comfortable rooms. Restaurant on the ground floor. €115.

Hotel Art Nouveau (F C4)
→ *Leibnizstr. 59,*
Charlottenburg
Tel. 327 74 40
www.hotelartnouveau.de
The rooms are decorated in a variety of colors: red, yellow and blue, which are inspired by Italian or Japanese paintings. This is a place where charm and comfort are of utmost importance. €116–176.

Hotel Askanischer Hof (F C4)
→ *Kurführstendamm 53,*
Charlottenburg; Tel. 881 80 33
www.askanischer-hof.de
A very elegant hotel with a Jugendstil entrance hall, beautiful classic German furniture, carpet and 16 spacious rooms overlooking the Ku'Damm. Ideal for a peaceful and restorative stay. €117–185.

Lux 11 (B D4)
→ *Rosa Luxemburg Str. 9–13,*
Mitte; Tel. 936 2800
www.lux11.com
A recent, wonderful design hotel in the thriving Mitte district. So fashionable it offers its own clothes boutique – Ulf Haines – and an Aveda salon and spa. Lux 11 is an apartment hotel, meaning the 48 rooms are well equipped with kitchen, pale stone bathrooms and modern, stylish furnishings, but there is no round-the-clock

attention from a front desk nor daily room service. Long white communal table for breakfast. Excellent restaurant. From €120.

Bleibtreu Berlin (F D4)
→ *Bleibtreustr. 31,*
Charlottenburg; Tel. 884 740
www.bleibtreu.com
Berlin's first design hotel manages to remain a hub of calm and tranquility despite its location between Ku'damm and Savigny Platz. The food is organic, you can get reflexology or acupuncture treatments and there is even an on-site florist to fully invigorate the senses. The 60 rooms are bright and eclectic, offering a good range of amenities. Restaurant, deli, bar and a very pleasant courtyard. €124–227.

Park Inn Berlin (C A2)
→ *Alexanderplatz, Mitte*
Tel. 23 89 43 33
www.parkinn-berlin.de

A match for Alexanderplatz, this gigantic hotel was originally built to welcome travelers from neighboring Communist countries visiting the former East Germany: 1,006 superbly equipped rooms and suites, and the most stunning panorama of Berlin since the picture windows of this enormous building were surmounted by a casino. From €129.

Hotel Riehmers Hofgarten (D B3)
→ *Yorckstr. 83, Kreuzberg*
Tel. 78 09 88 00
www.hotel-riehmers-hofgarten.de
Large, bright, well-equipped rooms elegantly decorated in a contemporary style. Excellent gastronomic restaurant, e.t.a. hoffmann, on the first floor. €129–145.

Arte Luise Kunsthotel (B A4)
→ *Luisenstr. 19, Mitte*
Tel. 28 44 80

The letters **(A, B, C...)** refer to the matching map. Letters on their own refer to the spread with useful addresses (restaurants, bars, stores). Letters followed by a star **(A★)** refer to the spread with a fold-out map and places to visit. The num

CITY PROFILE

■ Became the capital of Germany again in 1990
■ 3.8 million inhabitants
■ 341 square miles
■ 12 districts ■ 3 opera houses, 150 theaters

CLIMATE

■ Cold and snowy winter (32˚F), mild summer (63˚F). It can be very cold (5˚F) or very hot (95˚F).

STREET NUMBERS

Usually no odd- and even-numbered sides of the street: numbering is continuous and 'turns' at the end of the street.

1 Mitte
2 Friedrichshain-Kreuzberg
3 Pankow
4 Charlottenburg-Wilmersdorf
5 Spandau
6 Steglitz-Zehlendorf
7 Tempelhof-Schöneberg
8 Neukölln
9 Treptow-Köpenick
10 Marzahn-Hellersdorf
11 Lichtenberg
12 Reinickendorf

— The former Berlin Wall

BERLIN DISTRICTS

WWW.

→ *berlin-tourist-information.de*
Official website of the Berlin tourist office.
→ *berlinonline.de*
Current affairs, tourism, business.
→ *kulturprojekte-berlin.de/en*
Website of the association in charge of the city's big cultural projects.
Lists the must-see of Berlin culture (exhibitions, shows etc.).

Cybercafé
EasyInternetCafé (**F** E4)
→ *Kürfurstendamm 224*
Daily 6.30am–2am;
www.easyeverything.com

TOURIST INFO

Berlin-Tourismus-Marketing
→ *Tel. 25 00 25*
Mon-Fri 8am-7pm;
Sat-Sun 9am-6pm;
www.btm.de

Hauptbahnhof (**E** D1)
→ *Ground floor*
Daily 8am-10pm
Europa Center (**F** F3)
→ *Budapesterstr. 45*
Daily 10am-7pm (6pm Sun)
Brandenburger Tor (**A** B3)
→ *Pariser Platz*
Daily 10am-6pm
Ku'damm (**F** E3)
→ *Ku'Damm 21 Daily*
9.30am-8pm s(6pm Sat-Sun)

TELEPHONE

UK / USA to Berlin
→ *00/ 011 + 49 (Germany) + 30 (Berlin) + number*
Germany to Berlin
→ *(030) + number*
Berlin to Berlin
→ *Simply dial the number*
Berlin to UK / USA
→ *00 44 (UK) / 00 + 1 (USA) + number minus '0' for UK*
Emergency numbers
Fire service
→ *Tel. 112*
Police
→ *Tel. 110*

DIARY OF EVENTS

Public holidays
Jan 1; Good Friday;
Easter Monday; May 1 (Labor Day); Ascension Day;
Whit Monday; Oct 3 (national holiday); Dec 25–26
January
Lange Nacht der Museen
→ *Last Sat*
Many of Berlin's museums remain open until 2am.
February
Berlinale
→ *Second and third weeks;*
www.berlinale.de
International movie festival, Potsdamer Platz (**A** A4).
March
MaerzMusik
→ *Second week;*
www.maerzmusik.de
Experimental and contemporary music.
Festtage: ten days of classical music at the Staatsoper (**A** D2) and at the Philharmonic (**E** D4) opera houses.

April
Berlin Biennale
→ *April-May;*
www.berlinbiennale.de
Festival of contemporary art.
May
→ *May 1*
Radical and alternative demonstrations.
Theatertreffen
Festival of German-language drama;
www.berlinerfestspiele.de
Museuminselfestival
→ *May-Sep*
Movies, plays and concerts in the open air.
Karneval der Kulturen
→ *Three days over Pentecost;*
www.karneval-berlin.de
Multicultural parades in Kreuzberg (**D**).
June
Musikfesttag
→ *June 21*
Music festival.
Berlin Philharmonie
Waldbühne
Major open-air clas

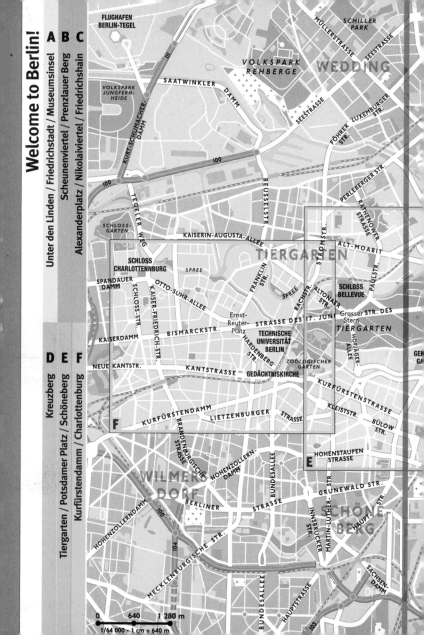

Welcome to Berlin!

FLUGHAFEN BERLIN-TEGEL

SCHILLER PARK

MÜLLERSTRASSE

SEESTRASSE

VOLKSPARK REHBERGE

WEDDING

VOLKSPARK JUNGFERNHEIDE

SAATWINKLER DAMM

SEESTRASSE

FÖHRER STR.

LUXEMBURGER STR.

KURT-SCHUMACHER-DAMM

TEGELER WEG

BEUSSELSTR.

100

PERLEBERGER STR.

RATHENOWER STRASSE

SCHLOSS-GARTEN

KAISERIN-AUGUSTA-ALLEE

STROMSTR.

ALT-MOABIT

TIERGARTEN

SCHLOSS CHARLOTTENBURG

SPREE

FRANKLIN STR.

SPREE

BACHSTR.

ALTONAER STR.

SCHLOSS BELLEVUE

PAULSTR.

SPANDAUER DAMM

SCHLOSS-STR.

OTTO-SUHR-ALLEE

KAISER-FRIEDRICH-STR.

Ernst-Reuter-Platz

STRASSE DES 17. JUNI

Grosser STR. DES Stern

TIERGARTEN

KAISERDAMM

BISMARCKSTR.

HARDENBERG STR.

TECHNISCHE UNIVERSITÄT BERLIN

HOFJÄGER-ALLEE

GEM GA

NEUE KANTSTR.

KANTSTRASSE

ZOOLOGISCHER GARTEN

GEDÄCHTNISKIRCHE

KURFÜRSTENSTRASSE

KLEISTSTR.

BÜLOW STR.

F

KURFÜRSTENDAMM

LIETZENBURGER STRASSE

BRANDENBURGISCHE STRASSE

HOHENSTAUFEN STRASSE

E

WILMERSDORF

HOHENZOLLERNDAMM

HOHENZOLLERN-DAMM

BERLINER STRASSE

BUNDESALLEE

GRUNEWALD STR.

SCHÖNEBERG

100

MARTIN-LUTHER-STR.

INNSBRUCKER STR.

HAUPTSTR.

701

MECKLENBURGISCHE STRASSE

BUNDESALLEE

HAUPTSTRASSE

SACHSEN-DAMM

103